Matter Mattering Matters: A Scienticity Reader

For Nanedi

THAT WHICH EXCITES
Kabelo Malatsie & Lantian Xie

A kick drum announces the song, a rhythmic tension that builds to a peak, dropped in your gut, forcing a move, a paroxysm, channelling air out through a whistle, a move that pours out in a ululation of improvised lives. A vigorous letting go of tensed up bodies. An ad-libbed liturgy of township youth expressed in unintelligible vocal registers somewhere between scream, celebration, and fearless unbundling. Shaking to chords and scratches, scratching and bending a molecular itch, a letting off of steam.

The body drops, each time anew, kinetic waves felt from afar pouring out of an inverted drum in cacophony.

Mrs. Mudzunga was a sangoma—people who have access to multiple timelines. Her son Samson is a South African artist whose performative practice had, for a short while, gained international visibility. He makes burial drums that are used in performances. These drums were collected.

One night, his late mother came to him in a dream and told him that she wanted one of the drums back. He immediately wrote to one of his collectors about this demand by his late mother—a demand which upset the laws of physics.

Stephon Alexander has formulated the universe as an instrument that constitutes itself through playing—makes itself matter (in meaning and form) through its mattering. The universe is an instrument playing itself. Just like the brain is an organ that names itself.

If You Don't Spin, You Don't Matter

Matter *of*, matter at hand; solid, liquid, material, thought, fabric, of the universe itself–in daily parlance; seeking determination of—resolutionary *what is the*—matter, as to locate, a problem, equated platonically with a geometric dislocation—as in *nothing* is-slash-even (the), (not odd)—unto which matter comes extempo; ulterior to location, spontaneously, without a discernible point of origin, or in excess of determinism. All that matters, all that glitters, earthly matters, family matters, matters of the heart, only thing that, relevant matters, pressing matters, matters left unsettled—affinities, significance.

Science science

Scientist friends lament about being co-opted into data analysis, farther and farther away from the lab, the field, the object, the phenomena—farther from the *real* science, the *Science science*; an insistent stutter, like the one instrumentalized when applied to words that conjoin persons to places, subjects to objects—*from* from—a performance of determination; determined in the act of determining.

Sometimes, when pressed on what is behind this Science Science stutter, scientists speak of the object of their lament; the lost object of science, describing it as *sound*—like how good science is *sound* science; or describing it as solid—like how *hard* science is *more* science than *soft* science. The lost object of science is a hard sound, like the criteria for proof; *soundness*, in which sound is a synonym for precise calibration. Precision proves itself so that something can be called *one* by the episteme.

International Rules-based System

The Haitian Revolution started in 1791, 233 rotations of the earth around the sun ago. Two rotations after that, the standard meter, made up of 100 centimetres—the one still used today—was ratified in Paris, in 1793, defined as the longitudinal distance from the equator to the North Pole, passing through Paris, divided by 10 million.

The meter was ratified amidst news of Haiti on fire, in revolt against the very same people who were busy legislating their own great domestic moral crisis into lore. Summer of '93 was a time of contradictions in Paris; a colony in revolution, the first of many, a signal for the end of an era. At the same time, news about a *new* world, to be inaugurated by a *new* way; more accessible, more universal, more *accurate* epistemic instruments—the meter and the kilogram; interfaces with which to instrumentalise mass & volume; the legislative instruments of a unifying, international, *rules*-based system.

If the 19th century's preoccupation was determinative and the 20th was interpretive, the 21st century has started off disruptive, in which the figure of the scientist as the inventor/inaugurator of the future at the centre of society has been displaced by the figure of the technologist, whose motive is to transform—instrumentalise–the world into resource (human, natural, etc), machines and instruments (financial instruments, war machines etc). In this, disruption is not so much exceeding what came before but a concentration of new codes of compliances into new instruments, new figurations, new calamitous views of the world, to which science comes as verifier of morality—an epistemic legislative body that underwrites a revolving executive body.

Integer Morality

An integer is an effect of an epistemological apparatus. An integer is an ontological unit, the most common example of the absolute. An integer draws its morality from a monological highground—an anatomical supremacy. Integer morality performs a discretizing, atomizing, literalizing action—a cornering of matter at an absolute threshold. Volume discretizes platonically as property. Motion discretizes platonically as time. In the production of an integer, matter has to be turned into an instrument. This instrumentalization does not operate on the insecurity of custodianship. It's operating on an extremist ideology that it is the word of god.

An integer's basic premise is that there is some truth to science; something more real, more reliable in it, in spite of the annihilatory limits it imposes on even its own terms. An integer is a definite article—stationary, discrete, snapped-to-grid, to which when force applied; motion as consequence. Instead, take motion as base—that things are always already in motion.

Intelligibility is a regime of sense-making. Newtonian science and algebra consider exponential functions but the limits they produce is to give an ontological definition to integers. But 1 is not 1 and everyone knows it. We don't even know what 1 is.

To conceive of an object moving in spacetime with no dust, no wind, no heat, no gravity, no curvature—theoretically, you have to eliminate any resistance to inertia; a child growing up or an apple rotting, in which transformation is reduced to loss.

We don't live in a euclidean world. Space curves. Matter, like subjectivity, is not singular. It performs multiple states simultaneously. Think of abundance. It's not building up towards something, some climactic anything, it's just abundance.

The morality of the lab is its sanitation but the perimeter of a body is always breached. Now, consider the hubris in drawing the perimeter of a billion-trillion bodies cycloidally connected, in which 1 can be a multiplicity of changing quantities, densities, weights, spheres, degrees—the measures of influence.

That which is unknown to the scientist is eulogised as the limit of science; the limit of knowledge, the edge of the universe—a higher not knowing. That which is unknown to the village scientist is regarded as the limit of exceptional circumstances, the limit of the village, the edge of ontology, esoteric, or admitted patronisingly, or abjectly as in *if you only knew.*

Jahannamiyya: Hell-way

Sylvia Wynter writes on Maxwell's demon—the nineteenth-century thermodynamics formulation of an entity, a force, an actant effortlessly opening and closing passages between A & B portions or coordinate functions—appearing *in* and also exceeding geometry, at will, moving in non-sequential, non-linear, extemporaneous, extra-planar, exponential terms, violating the idea that entropy must increase as time passes; that life itself violates determinism, trysts with gravity—that any old thought defies physics.

Katherine McKittrick speaks of "systems that cannot have a determined, or knowable, outcome", or what Einstein might have called "spooky".

Barry Harris speaks of "feeling the '*and*'", perhaps not unlike the '*yes, and*' in comedy; reading the sound underneath the keys—not mistaking the keys for the sound, in which a *key* can be construed as an appearance in geometry; an act, in which appearance can be construed as a contingent, partial interface for that which exceeds.

The contention with the scientifically legislated world is that what has been calculated or taken into account are things that are meant and imagined to be containable. What is uncontainable is construed as minor, or to be annihilated. But the world is not a minor percentage on a probability scale. It cannot be reduced to the production and management of calamity. It is abundant, dense, opaque, unimaginably possible, rife with material twists and turns and swings and swerves and spinoffs.

Whatever is significant in life does not have space—cannot be fulfilled in Euclidean terms. Matter is never not excitable. Picture a transmissible sociality, or, a cosmo-sociality, or, a socio-physics, to move and to be moved by that which excites.

We see the world in discursivity, in artistic practice, in autodidacts, in excitable, nondeterministic dispositions—as mattering practices as much as they are theoretical claims as much as they are indeterminate, expressive, lively, insubordinate.

Think of mattering in this way as a constitutive act, as an expressive act, as an act of thought spinning claims, wagers, riffs, dares. In discontinuous jumps, exponential, unprecedented, nondeterministic, defiant turns.

Meaning and form are not a priori. We make things matter and make things into matter by making them matter to our thoughts and lives.

Cartesian-Newtonian-Euclidean Computation Regime

An oceanographer overheard in a film remarked: "if you want to be right about shape, you have to lie about area. If you want to be right about area, then you have to lie about shape."

Picture a computation regime whose primary preoccupation is the ontological unity of things, staging truth as mathematical truth—two follows one, deterministically; two is more than one—twice more; that numbers, since made to be true, can be brought in contact with other numbers and that their interactions will be true, stable, repeatable. Think of literalism. It would not work without this premise.

If you search the same data long enough, you will find a false positive proof for any correlation—proof hacking, in which proof is the currency of confidence; absolute confidence is construed as truth; p-hacking comes to produce positive correlation through repeated application of the instrument.

The computation regime fantasises about performing repetitions in a state of inertia but inertia does not exist and so its power is met with constant resistance, leading to its brutal exercise of repetition.

The computation regime knows that no two turns of the wheel are the same and so it is dogged by probability, by sinusoidal *n* bodies, by a loss of empirical certainty.

The computation regime dreams of flow as an endless repetition without difference but that's not flow, that's just a wall clock floating around in the cosmos declaring exactitudes, deluded that the universe is operating on its time.

Proof is in the pudding and pie is in the sky

There was a story circulating some years ago around a paper written in some science journal by researchers at CERN hypothesizing on the incident of a pigeon who dropped a small piece of shit from mid-air at such a precise angle and velocity that it went into some ventilation shaft and hit some precise part inside that caused the whole collider to break—took months to fix.

The researchers hypothesized that since the collider was at the time running experiments on anti-matter and god particles, that the incident of the pigeon shitting on the collider may have been some effect of a force exerted on probability, backwards across spacetime, to prevent its own discovery.

Carceral Determinism

An old woman was accused of witchcraft and there were claims that she could fly in a loaf of bread or a 2kg bucket, at the time of this story yellow buckets were common. Questions of how shrinking portions were used for her to fit into the bread or bucket were all over. But these materials have no flying capacity so how could they fly? Was astral projection misunderstood as real flight? If so, why implicate the bread or bucket. The people who were accusing the woman of witchcraft believed that an old woman could fit into a loaf of bread or bucket and fly great distances. The accusation itself was proof, evidence works differently in this system. In a related case, christian prayer warriors believe that you intervene in the spiritual realm after midnight and before five, I remember 3am being an ideal time, at this time the world is full of demons and nefarious alchemy at play and the word, sound and moisture is the ultimate defence and cloaking device.

There is a word; gedankenexperiment, or *thought experiment*—another legacy instrument—whose ge- prefix signals a past tense, as in; prefigures the event, dislocates the thought from the event of thought, effects an observational discretization, formalises, ulteriorizes knowing as an owl's flight at dusk, a remote viewing as a prerequisite for knowledge of/on as split from phenomena in a deterministic causality.

Think of justice within a Newtonian framework—cause and action in an adversarial causality model whose outcome is resolutionary and inevitable, in which determinism is construed as a carceral force. A punitive determination is designed to account, to equate. It does this through the production of non-compliance through scrutiny, applied as observation or identification. Scrutiny is motivated to produce the object of its motivation. It is an adversarial model—in peer review, the peer is a referee.

A claimant addresses a courtroom in a language the court does not know, presenting an object the court is unable to admit. The claimant declares that if the object is to be seen, the world as it is known will be destroyed. The court stenographer turns to the judge and asks: "Your honour, how should I record this?"

J. Kameron Carter speaks of an alternative imagination of matter; an oratory ecstasy of speech outside of self.

Ashon Crawley speaks of social organisation signalled by shouting, speaking in tongues, noises, effects of the ecstatic, the anarchic—an excess of meaning beyond ontology; W.E.B. Du Bois' frenzy of sermonic discourse as a reconfiguring of matter itself.

The receiver of the act of determination receives it as recognition. Recognition generalises in scale–it performs a geometric consolidation. Visibility is a measured inclusion in a universe that is continuously and exponentially expanding. An integer morality tries to include parts of the universe that have not yet come into its viewfinder, but phenomena are not waiting to enter its radar. The viewfinder is in their domain. Think of any vital force. Think of affinity. It is nondeterministic, like when you think of someone and they hiccup.

No Fear of Time

A few rotations ago, a team of archeologists caught sight of a cave painting in Sulawesi, depicting a hunting scene from over forty-five thousand rotations ago, twenty thousand years older than the hunting scene on the walls of Lascaux. When asked what this meant for Indonesia in relation to the origin of modern humans, Basran Burhan—the archeologist leading the team—remarked that more than entering a new contender to claims on origin, what the Sulawesi cave painting makes possible is a rethinking of origin altogether, that seeing Sulawesi in relation to Lascaux was to see the world in a "boom of cognition".

Think of an *n* body problem—an excess of alignments, too many influences, too many gravitational forces to calculate. An *n* body problem's problem is its integer conceit, construed as a definite article; a body as an ontological unit. But bodies in centripetal orbit, orbit bodies that are themselves party to more orbits.

There is six times more gravity in the universe than all its mass and energy can account for. Think of phenomena thought in two ways; first, as a problem. Then, as a boom. Location demands coherence but coherence is simply a generalisation of scale.

Polar Bear, Cartesian Bear

A square demarcates a territory, draws a box for a cat. A circle narrows to a point, expands to a sphere of influence.

The old science pictured the universe as classical music, tightly composed, formalist, geometric, as in an orchestra where 1 body plays 1 instrument.

The new science happens as jazz goes to bebop—as philosophy comes into physics—playing music where everything is an instrument, where the universe is an instrument that plays itself.

98 rotations ago, in 1926, the International Federation of the National Standardizing Associations (I.S.A) was established, then later disbanded during the second World War, and then re-organized again under the current name, I.S.O., in 1946. With it came anthropology, coming to science in search of cryogenic technologies. "Even the name of the organization is standardized. The name, 'I.S.O' is not an acronym but was derived from the Greek word '*isos*' meaning 'equal'. The relation to standards is that if two objects meet the same standard, they *should* be equal."

In the 90s, the shape of ambition for enterprising uncles was ISO-compliance. The picture it conjured in the mind was a track—a world highway, a main line, with tributary tracks and on-ramps and off-shoots from which one could, through the act of compliance, receive admittance on the main-line track. The promise of this trajectorial alignment was to no longer reside at the smoking-gun, business-end of science where its premise breaks itself, at convenience—but instead, to move to its ergonomic handle, articulated in high fidelity to sustain its surface integrity.

There was a copy print shop everyone would visit, where the owner's daughter worked when she was off school, photoshopping ISO certifications one after another at lightning speed.

A trajectorial cohesion is a compliance with formalism. Think of a word disavowed or reconfigured with a prefix attachment—*de-, dis-, an-, non-, quantum-* —prefix swapping as reform work in which the root stays intact.

Intelligibility demands cohesion, demands stability—a platonic generalisation. But matter is excitable, not at rest. Picture life in excess of geometry, in excess of formalism; out of turn, not knowing one's place; indeterminate, swerving, twisting, shouting, contorting, jumping, happening all the time, out of nowhere, somewhere or the other.

Huey P. Newton theorised that "...contradiction is a ruling principle of the universe, and everything—phenomena, whether it's the physical world, or the biological world, or the social world, has its internal contradiction that gives motion to things—that internal strain."

Picture a worshipper on a faith-walk, enroute to their audience with god, from whom further blessings are bestowed upon those who have come from afar. This worshipper counts distance with—instead of footsteps—rotations of the wheels on their car. Picture this worshipper driving circles around their place of worship, taking meandering routes, spinning more blessing, rubber screeching against asphalt, exhaust billowing smoke as incense, cruising to worship.

Hortense Spillers speaks on *torque,* as " the place at which things begin to rotate again", of enthusiasm—*enthusiasmus*, in the Greek, as in; to be in a state of frenzy, or, to be inspired or possessed by god.

God likes me when I work but loves me when I sing

In music writer Jason King's liner notes on D'Angelo's Voodoo, ?uestlove relates: "The thing that really attracted me to D'Angelo's music was this inebriated execution thing that he had, which we both got from J. Dilla. Dilla would program his drums non-quantized...prior to *Voodoo*, I had been going through a period where I felt I had to prove to people that I was a machine: I had decided that my playing was going to be cold, you wouldn't be able to tell if I was a sample or not. I spent three years of just icing my presentation to a science where I was just a kind of super-metronome; I spent three years of trying to hide myself. Then, in walks D'Angelo, and he basically tells me: 'Yo, I need you to strip yourself of all that coldness and play human. I need you to play fucked up!' He wanted me to play as drunk and as slow and as dusted as I've ever played in my life...The first year of recording he would say: 'I need you to keep the pocket but don't drag behind me, but play a little crooked,' if that makes any sense whatsoever."

Raphael Saadiq spoke on this inebriated execution as performing with *the grease*, a "loose, way back in the pocket feel", a "rubber band feeling".

Tsshhhhhhh

Some decades back, there was a drought, the rivers were dry and the pipes were pulling air instead of water from underground, all we heard was tshhhhhh from the hissing pipes. ZCC church leaders went to the ruler to ask him for permission to perform a prayer for rain. They invited everyone to attend and join in prayer. The ruler agreed and the announcement was done on foot by blowing into the horn of a kudu. The day arrived and we all went to pray for rain by the river. It was an incredible sight of singing and praying on a hot day. A fluid dynamic was at play where breath was calling forth more moisture, collective breath was being given to receive more and more moisture, an exchange of gaseous state.

I remember digging a hole in the dry river sand and waiting for water to pool and slowly come up, waiting for the sediment to settle before collecting the clean water. Like low hanging clouds that are so low you could touch them, or as if you were bathing in the low glow of a fog, cold steam all around you...the strength of a rain maker is their ability to make mizzle and drizzles. They are to make rain that soaks the earth without erosion. It's rain you can dance in, cool off in.

A gardener nearby tends to azolla filiculoides, the fern which cooled this planet roughly fifty million years ago. It is a hotel for bacteria that can recombine nitrogen. From the balcony of an apartment block, a musician who sang about medupi a while back sees a rain cloud trailing behind the gardener, water pouring down from their very own garden hose out of the sky.

"THE HOUSE ALWAYS WINS"— HO PHEBELA? THE MATSHARARA PARADOX AND THE STORY THAT WON'T MATERIALISE—LOT 2 OF 7: LEKHAKHARI/MBILI/TWADIMA (DANGER-1100 [ELEPHANTS AND TORTOISE/0111[SECONDARY ELEPHANTS] 1101 [MAIN SHELLS])

George Mahashe

"Tjhokhothotjhokotho! Ṋaga Le a lla, le lla le lebeleḓe mphaḓalajana Khadi ya Jojo. La ṱhava, le khe phaḓalaja dikhadi ja Maake. Matsharara mulodi wa mutjho—e re Gong! ga re Gong! Ha ge a dzuha ge e nosi. Maḍimamullo a vina gosha ya mutjho ya hu phaḓalaja vadimu, e khe fhepa khetjhava. Nna mukhosi wa Khadi u vewa vyane? Re vea ga hu reḍa Lerumo. Yena, muṱhanga wa vadimu he a eme a e nosi – Khelume. Vahulu va re ha a ruhwe! U reḍwa ge mukhegulu, Thwadima kherube kha vomakhulu—a emelela a e nosi, va re ge Muloi! Ge Lemama re a mama Mashogojo a vamye. Ge Khadi e ḍiile mulodi wa ditjhela. E emelela vaana va yona. Ya phebela, Vakhalavye va sala ba khe zwevujisa hure va ladzwiswa mamabo a vadimu hu diahalaḓe ene?

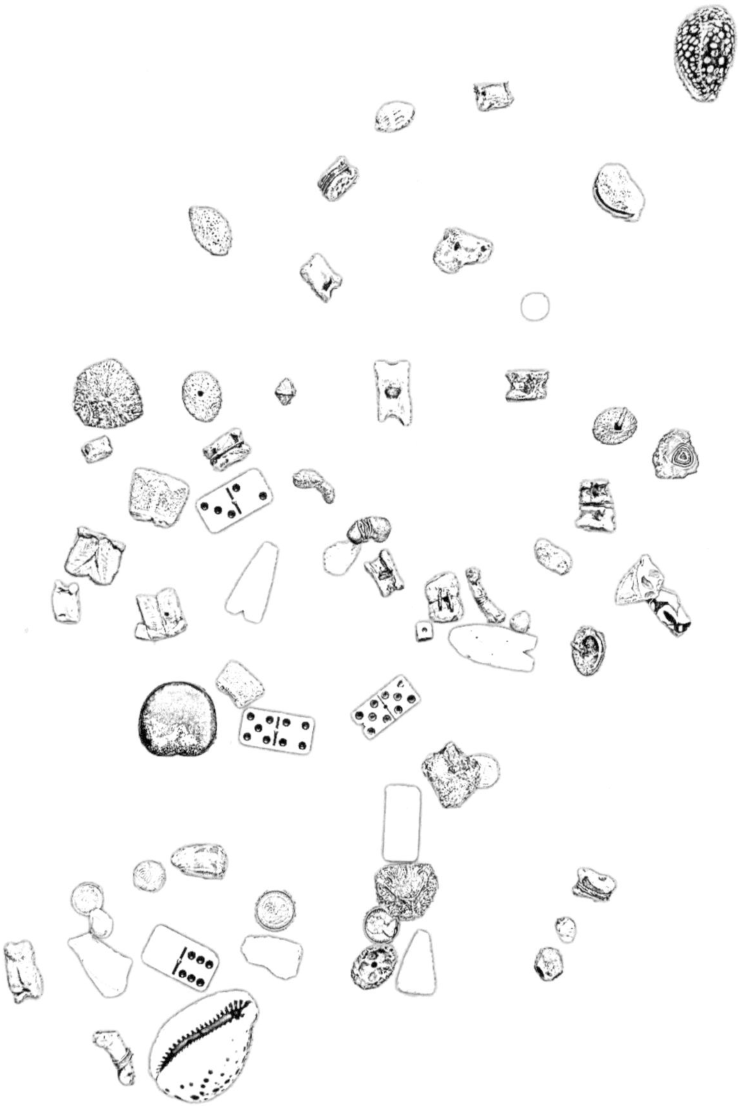

Vahale va re, muruba wa hu aṱhula ja Vadimu u vijwa ga ditjhela! Tjhokhothotjhokotho! Vujisane vakhegulu va ḓo le vinela gosha, va re! Khųba khuba re a khųbela! Zwereḍo ha zwe khuubye! Ge zwona muyemedi wa vathu khoroni. Ja re! [Diḍaula] Lerumo le le lebeleḓe Vakhalavye, diḓou ja hu vuya e e dikhudu. Di ya ṋeṋegeja, di khe ṱhallela vasimana va khe lega hu hasha ga Lumye (Lunḡwe)/Lulwe—. Va re! leetjha vaana va mba ba thakhale, va ḓo dia Khosi. Khosi ge Mukhalavye! U no levelela marumo a khe ṱhavana. Vujisa ṋaga le vakhegulu—phebelane khediguluho kha fractal! Ge kha hu phema Lerumo! Reḍane Masase! Khadi ge hu ḽesa Vakhalavye mamabo. Zwelalelo zwe vusa vadimu! Vakhegulu va re! Myana wa vaana ba mba u vusa vadimu!"

Binary	Character relation	Molawa 1 Lovedu	Molawa 2 English	Molawa/summary/remedy
1000	Old man up	Nkwame	burden carrier	Go belaela. unburdened, endurance, withstanding—father's father's spirit causing illness and problems—rakgadi to blow water to cool spirit. Affects sick animals. Unburdening of the heart and free will. (Calls for effort—phasa to encourage light.)
0100	Young man up	Selomi	imposter, stabbing, attacking.	Presence of spears—Motsamaya bošhego, mošegare re thusana le khosi (boloi by someone close). Illness of child related to lobola—Malome ga a phase.

0010	Old woman up	Twakadima/ Thwagadima, Takwala??	Quick short breath or panting (strike-through)	old woman sorcerer-di ka se thube-ngakatšhupwa in my case (father's sister's hoe = rakgadi)—for boloi, it must be coupled with Apron String. If the old woman is in the direction of the diviner (called lehlake), o kgona ka Baka la badimo—o rapalatša Manama = protected.
0001	Young woman up	Lunwe/ Thogane	we shall see. Short sharp sound (indeterminate)	lentha (clicking of the tongue). Illness in blood or stomach. Be on guard against enemies (phasa to ward off darkness or because of trouble ahead.)
0111	Old man down	Morarwana (Major), Nthakwana	fatigue, general upset. Small circle	ntwa lapeng—brothers fight on account of wives, mother's spirit tired, thuri—work of baloi. No journeys.
1011	Young man down	Moraro	Hewing, prosperity-digging or ploughing. Great circle	(moremogolo)—ntwa molokong—kgwara (ga go yo a ka go thušane)—thekga lapa—approach spirits yourself if sick. Rakgadi not far. If it falls three times, do not travel. Rakgadi to cut/track down bogodu ka leloko lapeng.
1101	Old woman down	Khadi, Mpululu—Mokholokwane.	ululation, lost property returned	Phasha pasha dibeke e ka se loke—run away wife not good.

1110	Young woman down	Lekhulela/ Makhulela/ Makulela/ Makgolela Lulwe or Nthekwane	bawling, crying noisily— I bawl about you	connected with the head of the lineage (Father's father's Spirit causing trouble through illness)— libation to ancestors mandatory. Head of the village must be approached. head of village to solve— be careful.
1111	All standing	Mpherefere	Tower, discord, continuous fighting- couple, storms (black and white colour. Fighting in the village)	Troublemakers to be driven out (small mistakes punished harshly). Batho ba ke dire dikgoši. If it falls continuously, then; issue between a couple via the wife— pherefere, go a baba.
0000	All down	Mohlakola	Sweeping. Loss of property or family member	darkness, pain, distraction, stripped naked, sickness and unhappiness—you work for nothing or fruitless labour— empty hand. Asking the question: what causes this?— Dark meat or dark medicine by boloi— calls for analysis (batho nthuseng le nyakele ngaka e alafe).
1100	Men up	Lekhakhari, mbili	Two stations, gristle— danger	Connected to the head of the family, good omen except when a woman is sick. Tough birth.
0011	Women up	Morarwana (minor??) ~~*Moraro??*~~ ~~*Vhakata*~~	Apron String, enfold	Not far, close to the gate (danger??)— to restrain (enemy ineffective). Recovery from illness except for women.

1010	Elders up	Muhuluri -Go potla	bellow	Happiness (mom's spirits). Should be associated with difficult concepts–often avoided by citing that it's above my pay grade
0101	Youth up	Bahlapi-kwena; zwimbili	twins. Ditto, Crocodile in the pool, tramper	All good. No illness. Good omen—strangers not allowed in village—boloi from outside. Bothata boa di pheo goba sebudulelo. "Lesogana le tla aga motse le kgarebe ye e le ratago."
1001	Old men and young woman up	Tshitangu	Shoe?? Joint, take shield when travelling	Father to end the quarrel between brothers.
0110	Old woman and grandson up	Maropeng, mogogana	tugging/ tagging, mother's womb	Problems caused by mother's spirit—where the umbilical cord is buried.
0010 +0011 or 1011			Restrain from using sorcery (boloi)	See new note book for comprehensive combinations and effects.

Six-sided cube dice according to my father:
1-Marumbi (Maropeng) = mother's spirit
2-Khadi/mokholokwane
3-lungwe (ga go be bwalo—amen)
4-mohlakola = seek a diviner
5-moferefere (mohlakola by Mamojela)
6-nkwame (Baloi by Mamojela)

The praise verse formulation we opened with above is associated with the 16-bit coding of the Makhalaka four-tablet divination system—Ditaola tša Bakhalabwe. Ditaola are also referred to as dithaku, the dice, or divination bones. In my bag [sekhwama] I have:

- Ditlou set (elephants, seniors) to which the name Makhalaka refers. These are commonly made out of ivory (sometimes substituted with dominos), they denote space and experience and are read first;
- Dikhudu (tortoise, juniors), related to time and are read last to confirm the elephants;
- Dikolobe (wild pigs, MaNgaga (doctors and experts)) who diagnose the nature of relations as positive or negative as well as the inbetweens;
- Lewatle (sea shells) which speak to all things eternal;
- Tshelete (monetary coins) that speak to ideas of money/wealth/means/luck;
- Naga/leshago (pottery and stone shards)—these speak to matters of land, particularly stolen land;

Sssaaaaaaaaaaaaaaaaaaaaaaaa! [Nani's contribution]

These do not include all the other mechanics of the divination bone sets used in the lot presented here—spanning glass, wood, plant matter, assorted animal joint bones (hence the name Marapo meaning bones) and metals. For our purposes, I am restricting the discussion of elements in the lot to the 16-bit Makhalaka-style sets because these are adequate to diagnose all possibilities if cast 7 times, as we have done for this text. Similarly, in each lot there are six 16-bit combinations that are also confirmed or tempered by the other elements of the lot not discussed here. So, in essence, casting one lot would be adequate for a reading that dips into the '7 pools' at the same time. [The seven pools denote 7 dimensions of reality needed to locate.] Effectively, we are talking about a lake—where half-men meet once a month on a full moon to adjudicate the fates of their children and Mahosi.

These understandings and imaginings build on/with my father Obie Stanley Mmolawa's notebooks, passed on to me when I began to work seriously with several sets of divinatory bones encountered over the last 7 years. I draw on three different notebooks compiled at different stages of his life. They aggregate published accounts and studies from missionaries, anthropologists, as well as from oral histories collected during the time he spent consulting healers and preparing for his own initiation. The table shared is not complete and is undergoing corrections and revisions as I learn more.

A teacher sits at another's feet for thousands of years. Then, a missionary sits at the feet of that teacher. My father joined that procession, and now I must sit at the foot of a book and an experienced diviner in a myriad of pairs [teacher/student, father/son, mother/daughter, uncle/nephew, grandmother/grandson, ngaga/missionary, samaritan/wandering-meteorite/land-dream/social media, and so the procession goes].

At the core of this divinatory system (Makhalaka and the other elements not discussed) there are accepted assumptions about relations between the feminine and the masculine construed not as gender but rather as a set of complementary relations in a broad spectrum. Southern African oral tradition maintains that Motho (an aspect of Ubuntu) is made up of two inextinguishable entities often referred to as half-men. The right being tona while the left being tshadi. Tona being the force that upon dispersal (death) travels to the source and remains there to teach others while tshadi is the force that returns to the source and immediately returns to guard its offspring (not necessarily biological). For Motho (sometimes referred to as HlogoTšhweu—the Whitehead or wiseone) to exist, these two forces should exist as one unit in our time and space—I am sure there are parallels to this in the world of electrons and so forth. So far, this is quite standard to general biology—I will not attempt to transcend this language. Within the broader system, there are pieces that are not paired.

In addition, each die (divinatory bones are sometimes referred to as the 'dice') in the divinatory system can fall on two, three, four, or six sides; these falls can be passive, active, defensive, digging, or attacking. If it has two sides, it's either silent or speaking, getting more complex as the number of sides increases. Then, on top of that, there is direction they are acting upon; either coming or going or so forth.

When one casts the lot, it is customary to praise the fall in a poem that names the lot and emphasises the different positions and relations of the dice as they rest. There are 16 standard poems (the bits), which can be improvised into 64 variations within the system's interpretive catalogue as one casts more lots. The poems are a formula in the sense that it is composed, specifically designed, and impresses upon ingenuity—a composition that assesses a position/location. In our case with the dominance of the fall of Selome, one can say that the approach is not to accuse but to praise the keepers of the spear—Naka—baana ba lela. It is a praise that can speak ill and flatter at the same time.

Tjhokhothotjhokotho! Ṋaga e levile vadimu va vommahu! Vona va thekhile lerumo la maṱhadze. Va re, Marubini ge mo re vuyelaho! Nḓo ya mukhalavye ge lewaḓe! Le fhelehejwa ge muthu a fheleleḓe! Nna re vona Ḓou ya Nkwame e jerwe/mejijwe ge khadi ya lewaḓe! Thwahadima Le Nkwame va ya aba!—va aba ga lendzwi ga lethihi. Ha mukhalavye a edzeḓe, mukhegulu u eme ga mauḓu. He mukhegulu a edzeḓe mukhalavye u eme ga mauḓu–hu ga se we Mpherefhere!

Within my divination set there are two houses. We have the house that's the principle of all the dice (the leopard shell is the bigger one of the two), then we have an active house representing a place for airing out grievances—Kgoro (the cobra head cowrie is the smaller one). A peculiar thing about the seven lots cast for this reading is that on the first casting of the lot, the big house that is upside down swallows the old man [elephant], effectively becoming human [made up of two halves]. In this way, you cannot read the old man without reading the old woman. The old man has been swallowed by the sea, concealing his voice and putting them in perpetual disagreement or in a state where they never speak at the same time (that is, speaking in unison or in dialogical concert). When the old woman is active, the old man is sleeping, so they speak with one voice. This is particularly interesting because, as noted earlier, Motho—to be human—is to be one part male one part female as the two dice have now become—HlogoTšweu. So we can conclude that these lots are presided over by a wiseone. This invites a story around how the Mudjadji dynasty begins:

Kgale kgale lefase le sa tapunya, when all the Mambo's sons [spears] were fighting for the throne, the old ruler approached his daughters and offered them a chance to become the next ruler, complaining that while the young men are fighting, an enemy will take over the land. One of the daughters agreed and a house was built where the young woman communed with the old man. From that house, a new daughter was born and the throne was secured.

In another rendition of the story, one of the Mambo's daughters asks a diviner to sneak her into the monthly sitting of her ancestors; the Whiteheads. The diviner offers to give her something that will make her invisible to them and warns her that only her voice can betray her presence to them. She eavesdrops in on a discussion on what it takes to rule. She emerges from the sitting as the most powerful doctor in the land, Mudjadji—Mokhalabwe. But that is not the story we need...

Within this lot, at the edge of this drama, there is the presence of a unified whole in the form of a marriage of the sea and the elephant. It reads: Canopus points to your mother's lineage, reminding us that the solution lies in the family ruins, where the family's umbilical cords are buried. The grave of the old man is the ocean. There they speak as one complete being. The house that is passive, shielding the old man, points directly to an aggrieved young woman.

There is a complaint—a grievance.

Lekhakhari!—vuilo—happiness, because a grave is both ill omen and redemptive place, this is where you find renewal—1100–danger, confusion... khavu ya mullo—fire!

When I was a child, a diviner I knew died. After the funeral, the family opened the trunk where the diviner's things were kept and found that their Ditaola had spontaneously combusted. There was no one to receive them. This force is called magic by some but I suggest it is something to do with an inextinguishable energy (debatable as some maintain that the energy disappears when you stop interfacing with it, placing its logic as subject to engagement). Their energy had dispersed, waiting for the rightful person to emerge. Some call this fire Vutha, a rampant fire that starts out of nowhere.

I have come across a similar story of a fire that engulfed an archive of Southern African material culture in a university ethnographic museum in 1931. The rumour goes that some material was kept there beyond its term. The "custodians" were advised to return that material by ordinary people who knew of their nature but didn't have the authority to insist. Then, there was a fire that burned down almost the entire building complex.

When there is no clear successor, sometimes there is a fire. The material escapes, appears and reappears until it finds someone that can hold it. Ditaola di fiwa ke badimo, if it's not meant for you, it will not stay.

A house was in the process of promoting a spear. They searched the graves, and informed Nkwame, the old man of the house, that they had found muruga-wa-khosi, the insignia worn by the king. They asked him to come and verify it. The old man did not go. He did not want to encounter what they had found. They thought they had found beads but they were actually in the proximity of a dangerous entity—Sebata! Bakhegulo ba re Kgomo!

In some places, one is only a ruler because they are accompanied on their body by something fierce, something unimaginable. Usually, it is associated with danger or an incapacity to be in its presence. One cannot survive to be in its presence so being ruler is to withstand the company of such an entity.

Tjhokhothotjhokotho! Thwahadima u emelela a e nosi va re ge Muloi! Ge Lemama re a mama Mashogojo a vamye—Ge khadi e diile mulodi wa ditjhela—Ya phebela, Vakhalavye va sala ba khe zwevujisa hure va ladzwiswa mamabo a vadimu hu diahalade ene? Tjhokhothotjhokotho! Khuba khuba re a khubela! Zweredo a zwe khuubwye! Ge zwona muyemedi wa vathu khoroni.

In most cases, it has been said that this utterance is to distract the client while the diviner thinks, but to a degree it is an utterance of reality. So you praise the dice, and you speak on their behalf. You invoke the interpretive catalogue in perpetual hyperlinks until the appropriate formulation resonates.

The diviner speaks and knows. The divinatory bones stand as a placeholder for them to order their thoughts and intuition. Ditaola is not grammar but the device that gives grammar and controls the flow. In some practices, there are no dice involved—the diviner just tells you. Ditaola also renders objects, spatialises the conversation so everyone has something to hold on to. Hence the formulae, when uttered, distracts. In most cases the diviner prior to the divination would have already had a dream in which the complexities of the situation to be elaborated and addressed are unfolded. The consulting party themselves would have also had dreams. So the divinatory bones, and their associated interpretive catalogue, in this case helps both parties to hold on to a rather complex image that manifests as a dream.

Wim van Binsbergen notes that; "The divination apparatus [the 16 bit and interpretive catalogue] is a machine for generating an unlimited number of different stories: stories at least one of which contains meaning and revelation for the client." The training, or *play* if you will, is a way to learn the formulas towards a greater ability to spin stories. For the work, as ngaga Mamojela notes, 'ke go di abela'—to speak on their behalf, not to recite but to fabulate. To tailor a story unique to the client. A story that facilitates progress as the MotshwaraMarabo does. "To spin stories of ever greater depth, relevance, and drama—a directly applicable poem written under specific reference to the client's life and problems." To quote an anthropologist, Binsbergen and the utility of his text is to reinforce the idea that he was anticipated (admitted into the procession of sittings and teachings). Like Ditaola, he was used as a carrier of the interpretive catalogue for expanding the computer that Ditaola are.

It is also worth noting that what Binsbergen identified as the one critique of this divination system is that it neglects the effects of modernity and capitalism by redirecting to issues of ancestors and their protection (or lack thereof). The divinatory system's lack of emphasis on modernity and capitalism speaks to their inconsequential nature in the grand scheme of things—maintaining one's humanity by not being distracted by fleeting conditions. This lack of emphasis speaks to the predictability of colonialism and is therefore not without remedy. It does not warrant deviation from our path—di a buda.

As Sophie Oluwole points out in relation to Ifa (and by extension Khelovedu), it is not a religion/science but a storage mechanism stored via binary code. The binary code (not conceived in opposite but complementary) is a bit-system of organising the stories contained in Ifa. It is kept at the level of ancient philosophy before the division in favour of disciplines. As such, the 7 stories (resulting from the 7 lots in our case) are consistent with the binary logic of computational science—1s and 0s—and its desire to house all that is knowable. As such, the divinatory system is at once a theory and practice of knowing what can be known by way of reciting what has been known before and applying critical literacies and intuition (in the present) to guide oneself through the vastness that is the known (and known to be unknowable). It is also a meditation of what are stories and analogies, what they do (or interface) to observed reality. With my specific divination set, which houses several types of divination systems (both local and drawn from Central and West Africa), the binary system is amplified in all directions to facilitate a broad starting point to generate a story. For example, you can cast the full lot with a broad question, and realise the question is mathematical or domestic, and then focus on that for the succeeding lots. In this case, there are two types of questions—*direct* and *if then*.

As an "archaic" form, it is confined to the heyday of anthropology and archaeology as they once attempted to understand how the "other" accounted for our broader existence and susceptibility to forces beyond our control or imagination. Often, it is seen as an account of culture and ritual but even within the limited scope of anthropology, it was recognised for its capacity to reconcile aspects of lived and perceived realities in ingenious ways. As a living practice, it is persistent, passed on from teacher to student (through linkages beyond kinship) via tuition/instruction in the flesh. And when teacher and student do not occupy the same spacetime, it is passed on through dreams and persistent real-life material problems that engage the relevant research practices that lead to a similar knowing. It is also sought after by those excluded from it by lineage, although the very impulse of seeking it and practising its curious methods automatically draws one as a student, rendering the process technically open to anyone that applies the time.

The aim for this text is to lodge (dare) the practice of Ditaola [Dithlaku] within the discussion of European science's epistemic instruments and the broader question of encountering/understanding reality and its materialities (from domestic dispute to drought, to predicting the next supernova). This would be an extension of the practice from its current location as "archaic" wisdom into a contemporary practice that refuses to yield to the weight of coloniality—its versions of knowing, its practices of observing and rendering phenomena.

My own interest in the practice, aside from it being mandated by some ideas of lineage (maybe we might delve there later), is on account of their role as both practical and theoretical instruments, at once being able to measure direct input from contact or proximity (immersion in khediba (type of isolated water source)) to measuring what I can only describe as cosmic radiation in an effort to predict the death of Mahosi (tied to objects like meteorites). It also acts as indexical signs that are able to deploy metaphor and images to compute hypothetical scenarios like what social phenomena might be causing your clinical depression.

For the purose of this text, it is more important to go through its motions as I understand them, to release the results from the expectations of this spacetime into that of the half-men and the Whiteheads—HlogoTšheu. To cast the lot and see, is to enact without knowing. Ditaola are a series of hypertexts read through intuition or a knowledge of how certain situations lead to each other. It's a way of keeping in play a delicate idea, a practice whose key is to keep practising. In other words, this text is a wager.

Note on orthography: retaining the dilemma of an unresolved conundrum of KhiLovedu [Khelobedu] and SePedi (as they struggle to untangle from the colonial designation of KhiLovedu as a dialect of SePedi), is to bring to bear the type of manoeuvres one has to make to surface a story that transcends the designation of Lovedu or Pedi. As such, I retain the mixed vocabulary of SePedi, SeSotho sa Leboa and KhiLovedu, with the inconsistencies brought on by a changing KhiLovedu orthography. However, the poetic verses are rendered in Khelovedu proper, as translated by Khutatso Seshayi using the proposed orthography developed in the context of translating the Hebrew bible.

FROM INFRA EARTH TO EXTRA SKY
Aarti Sunder

The cover of *thanthiram, manthiram, yanthiram* says that the motto of the book is public wellbeing and inside on every other page is written that no part of this must be used for unethical purposes. The book expands with great detail and instruction on how to perform a wide set of potential desirables—from how to stupefy a nation, to how to have someone fall in love with you, to how to get rid of unemployment, to how to ward off fear of snakes. Elaborate and meticulous procedures specify actions that can help accomplish each of these desires, with no comment on the ethical veracity of any of these inclinations. The book also states that there are three stages to making incantational cosmic drawings as a practice, one of the most important ways to attain said desirables: one is to draw on paper which is least effective, the next is to make markings on a coconut and this considerably increases effectiveness and finally, drawings on a copper plate are most effective.

The most curious part of the road to fulfilling desires is not about whether the person is any closer to achieving what they want, but in the potential that something *may* take place. Hope is attached to these drawings, rendering them as powerful images of predisposition and determination. However, they are also a very specific type of unknowing, in which the scales of what account for success are tipping and not tipping at the same time, are always taking place and perhaps not as well.

So a few months ago, when I approached a crowd of people on Thiruvanmiyur beach who were excitedly and warily speaking over each other while pointing to a coconut, I was well aware of what was at stake. A coconut was covered with nails at equidistance, in circular formation and at its centre was a geometric drawing. Bits of hair were wound around each nail. The location around every nail was blackish as if burnt. The question being asked there amongst the people congregated around it was not *what* this coconut was meant to accomplish, or *what* kind of desire it was meant to fulfill. Instead there was a deep clarity of what sort of action should take place next, now that the coconut's presence was fact: *what should be done to it?* It could not be left as it was since it was impossible to tell what the nature of desire it was meant to fulfill was. What if a child were to touch it? What if it accidentally altered the course of an innocent onlooker's life? What if, what if, what if? (In fact, the mere act of photographing the coconut sent my mother into worried exasperation.) Normally, coconuts are broken open as an offering, or buried somewhere specific (there are rules to this too) but this one with its nails, drawings and hair was left intact and cast away into the sea. Leaving the coconut intact was a signal to its potency (which too has its rules).

Casting such an object into the sea is accepted practice, where the sea functions like the universe that is open, and wide, and unknown. A cast object goes *away somewhere*, anywhere, nowhere into the nether. The coconut was not meant to return as object, it was meant to return as potential that would produce a specific effect to human sociability. Except it had returned and this return was out of character and so had brought with it a flurry of caution and excitement. Chaos ensued only when the object itself returned as an undisturbed and detectable whole—the drawing on it was intact as were the nails and the hair.

Spectrum

So when in 2023 the Indian Telecom minister Ashwini Vaishnav stated in an explainer-addendum to the new telecommunication bill to be tabled in parliament, that *spectrum is similar to atma (soul), which is ajar (forever open), amar (forever youthful)... Like atma, spectrum too does not have any physical form, yet it is omnipresent,* he was certain that his words would not go unheard. His goal was to explain the word spectrum in a way he thought was easiest to understand for the Indian public. He found that if he were to turn the very idea of the abstract and concrete on its head and attempt to link the two, he would be successful.

The word spectrum can mean a few things—wavelengths (eg. light, colour) or particles of electromagnetic radiation that is emitted or absorbed by an object. It is also a non-specific set of values across a continuum, or a range of a type of thing. As a word it is open enough, vague enough, and importantly precise enough to allow for different interpretations to seep through, to always retain its original meaning and to see-saw between abstraction and concreteness. The soul on the other hand is a little tricky—it is immaterial, can be understood as energy, as within the real or the immortal, as something with intensity, as an ungraspable but life-giving or life-taking principle. The soul has great spiritual and religious undertones to it, leading our understanding of it to constantly shift between abstraction, further abstraction and concreteness. Both concepts exist regardless of location, time, space, language, caste, creed. Both have a relationship to the ether—the air, that which fills space, fills bodies, at once refers to airwaves and also the heavens. Ether is the sky, is hyperbole, is a hypothetical, but it is also stillness, or the very thing that makes all movement, and all life possible. Ether is—in addition—also a clear inflammable liquid that was once used as an inhalant anesthetic. In short, like spectrum and soul, it is a thing but it is also not a thing.

While the soul and spectrum may not have a direct co-relation, their relationship with adjacent concepts, like ether, made them unusually perfect words to use interchangeably, in place of each other and in conjunction with each other. Thus the relationship between spectrum as a technological device and the cosmic was made evident, while the soul (as always already cosmic) could be seen as a measurable, trade-able, physical entity. This abstract-concrete interchangeable link surpasses all life forms because all living things allegedly have souls. It also transverses all space, since it is there that we find wavelengths and particles, right from deepest parts of the earth all the way into the highest parts of the sky.

Ashwini Vaishnav continues, *"The nature of spectrum is different from the rest of the assets. Spectrum is an asset which does not deteriorate. If it is with X and tomorrow with Y, then there will be no difference in it while transiting from X to Y. It was not consumed. There was no loss in it...Unlike other assets, spectrum is a totally inexhaustible asset. Since the time the universe was created, the spectrum has remained intact. That's something very unique to telecom as an asset class."*

Even though he used the word spectrum in the passage above, he could have just as easily used the word soul. Spectrum as a technological asset and soul as an intangible quasi spiritual asset here see-saw between the concrete and the abstract; they are always changing phenomena and (according to him) are meant to be understood as that. Spectrum is made to simultaneously assume the stature of a proper noun, containing agency. However, rather than understanding technology as infrastructural and part of our everyday life, spectrum is rendered—as technology and as infrastructure—further into abstraction. *The spectrum is like the soul—* if the Soul is abstract, then so is Spectrum. But if the Soul is concrete, then so is Spectrum.

He finishes, "*Keeping that in mind, the entire focus is that there should be continuity of services...*"

If we were to remind ourselves that all of this is being said within the context of a telecommunication bill replacing the 1885 Indian Telegraph Act and the Indian Wireless Telegraphy Act of 1933, then we begin to understand the power of storytelling, mythologising and abstraction. The new Bill deals with allocation, administration, introduces high levels of surveillance, infrastructure building, services and private-public partnerships. Importantly, the Bill also grants the central government total authority to take control of, manage, and suspend all telecommunication services and networks at will. In 2023, Manipur saw more than 200 days of Internet shutdown. In 2019, Kashmir saw a total communication blackout for over 500 days, well into 2021.

Spectrum, like the soul, morphs into a service that is to be managed, allocated, mined and traded by the same entity (the State) that offers new parallels, storylines, definitions and associations. Spectrum, like the soul, is now a concrete; sans abstraction since it is a product of capital and government. The Cosmic becomes governmental as does the communication between the infra earth and the extra sky; it is regulated and mended, oiled, and made to churn, produce, align with central governments or state governments. In short, the cosmic is always already structural.

The drawings on the coconut function as regulatory, as social, as spiritual but also are shadowed, mysterious and impalpable. It leaves a space parallel to that of the spectral utterances that attempt to govern the nation-state-soul in a blatant game of purity and impurity. If the spectral utterances are orthodox and pure, the edgy scratchings on the coconut are seen as an impure vessel of extreme potency, meant to harness the forbidden and the dangerous.

They are analogous; one cannot exist without the other, sometimes clashing on the cosmic battlefield and sometimes coiled around each other inseparably.

The cosmic is not fact, but exists when it is explicitly *looked for* almost like something that needs to be willed into belief every time, similar to making a plan or a plot on a graph. The cosmic and our relationship with it needs to be constructed, altered, bettered, retold, discarded and remade each time we consider it seriously. It is opaque and transparent, networked and multipolar, drawing connections between in the infra earth and the extra sky. It performs and signifies, allowing for many characters to exist within it. The cosmic lays its foundations in epistemic purity but easily departs from it, moving away from the sensible dimension into knowledge of incomplete actors. Certainties of science can be pulled apart here so that truth and surety can be called into question. The spinner of the best tale takes precedence, the greatest inventor's voice is the loudest and the most daring imaginations travel the farthest.

Like the coconut and Ashwini Vaishnav's formulations of the Soul and Spectrum, many objects, and concepts, and thoughts travel into the furthest depths of the sky and into the cosmos. Some are pulled back with force and made to transform its surroundings. Some return to transform in a non-deterministic manner, as a *something else*, as a shape-shifting potential that can reweave an existing story. They collide and implode and chronicle *differently*, flipping narratives on their heads. It is in this *something else that flips differently* that new and old sociabilities exist.

The exterior (the place where thought-object-forms travel to) exists as a place to linger, but not remain indefinitely. The exterior is not a place to escape to, to run away to, but a place to reground ideas so that the inevitable return can occur. It is similar to the telescope—a profoundly remarkable visual prosthesis that has altered our understanding of the 'outside', scale, time as well as the way in which we construct images of different spaces in our minds. With the telescope we begin to see further, and further away. We get transported into the unseen and return instantly, see the world 'outside of' and linger there for a while. We send telescopes into the unseen outside. We await the information that always returns to us with the potential to alter how we read the world.

Sunny

In 2022, I met Sunny in Sharjah where I was visiting the aquarium with a friend, and he was one of the caretakers there. After Sunny saw us pointing to a few fish, he came over and told us that most of the fish there were sick. They were anxiety-ridden and almost manic in their frustration. One of his jobs was to enter the aquarium every morning and collect dead fish from the concrete floor. They would jump out every now and then thinking the waters outside were calmer and less treacherous. He saw himself in the same world as he did the fish and proclaimed that he too is like them. If the fish were trying to escape to a place more still and calm, so was he. In collecting and storing the carcasses of the dead, Sunny was certain their death would not be in vain. That they would return one day, and when they did, they would transform the ways in which we make society.

Smoke

In 1903, the SPCA, Thomas Edison's Edison Studios, an amusement park architect, and a circus master came together to perform the public execution of an elephant from India that had traveled to the US. Topsy was an abused and ungovernable elephant who had killed three men in the months prior to her execution.

While the architect of the amusement park she was in and the circus master for whom Topsy was forced to perform had taken the decision to kill Topsy, the SPCA insisted the murder be conducted 'humanely'; Topsy was electrocuted rather than hung. She was first served with cyanide-laced carrots and then was meant to stand on two copper plates through which 6600 volts of alternating current (AC) was passed. Thomas Edison was at the time set on winning the war of currents against Tesla's alternating current. So he sent 6600 volts of alternating current through Topsy's body to show that it was dangerous enough to kill such a large elephant. His hope in this public display was that the industry standard would shift to direct current (DC) which he claimed was less dangerous. The demonstration of alternating current at Coney Island was meant to see an easy 1000 people cheering for Topsy's death at 25 cents a head.

An article that appeared after her death stated that all the power from Coney Island, save for trolley cars required for transporting people, was rerouted towards Topsy. 'During the few seconds of her electrocution her tail shot out, and her trunk was tied in knots. A shiver ran over her frame and then all was still...she died instantly. For ten seconds the current was kept on. When it was turned off, the dead body of the huge beast swung gently for a second or two and then sank slowly to the ground.' She was strangled so that any accidental life remaining in Topsy might be snuffed out. The theatrics of the situation was not to be missed, 'four huge streamers advertising the enterprise were strung from the side of the platform.'

During this time, the electric chair was gaining traction as the most acceptable form of decreed death. 1890 saw the electric chair's first victim. 1000 volts of AC was passed through William Kemmler's body for 17 seconds after which he was pronounced dead. However, he was still breathing; the electricity had only rendered him unconscious. He was finally electrocuted again with 2000 volts ensuring his death. Execution by electricity was advocated for as progressive, bloodless and cruelty free; the magical sublime of the electric charge contained the supernatural and mysterious ability to conduct a painless evaporation of life. In reality however, witnesses at the time recounted scenes of horror: the roasting of a body along with bursting blood vessels, burning flesh, and erupting eyes. While Edison himself was against capital punishment, he still believed that electrocution by AC would ultimately lead to its stigmatization. Prior to Topsy, Edison was involved with the electric killing of many stray dogs, horses and cattle—to make the point of AC's danger for public usage.

Topsy is silent in the video of her death. But what was she listening to? And what remains of her today?

Topsy was never a mute actor in the by-lanes of a forgotten time or a milestone of technological prowess. She is instead a fulcrum of modernity in her determination to attempt life despite the pathological abuse she received and her lingering presence in the history of technology. Her anger and resolve to be ungovernable is evident in her final moments. The elephant is in the air, not in the room as most would think. The elephant is instead everywhere. The dissipating smoke from Topsy's feet floods the cosmic mass in its anger and its eloquence.

In the one minute video that Edwin Porter recorded on behalf of Edison Studios, we see Topsy being led through a crowd of people in what seems to be a construction site, where chimneys smoke, where people are hard at work, where the armature of buildings are laid bare. Topsy walks through these bustling half-buildings and a 200-strong cheering crowd. In the 20th second of the video, there is a cut in the frame; Topsy had apparently refused to walk to her death, to her designated spot on top of four prepared copper plates and electrodes that were connected to a generator. The copper plates and electrodes needed to be brought to her instead. While the cut is instantaneous, the break in the recording itself continues after 1.5 hours: we see Topsy strapped-in and remaining still. Soon, smoke starts to appear from Topsy's four feet that engulfs her in a haze; she falls and her body assumes a baked stiffness. One copper plate clings to her back left foot. Through a cloud of smoke, we see that Topsy is now dead.

The white transparency of the smoke dissipates slowly. The smoke spreads, thinning out until we don't see it anymore, evenly spreading itself like a cover over land and sea forms alike; the smoke itself, ripe with the smell of burning flesh, is untamed, quiet and animalistic; Topsy is now gaseous and her grip on modernity is uneasy and inescapable, permeating very crevice—smoking batteries, connecting cables, short circuits, burning circuits, heat surging wires, melting insulation and scorching flesh. Smoking fire carries water vapour and carbon particles capable of conducting electricity in heavy concentrations. Smoke from burning flesh is acrid, charcoal-like, and with a lasting stench. Loose wires, damaged connections, faulty grounds; the combustion of materiality into a smoky form that helps it to dissipate.

Topsy's smoke, much like a cloud or *the* cloud, see-saws between the material and the immaterial. Like the cloud is the non-object that moves through the sky—impossible to touch, but possible to feel—is imbued with water vapour, crystals, and is an eternally shape-shifting abstract form. So is smoke, and specifically Topsy's smoke. If the cloud was the chosen word to refer to an amorphous network of telephonic, cable, and internet spools, it was precisely because it could be considered everywhere and nowhere simultaneously. Topsy's smoke too must be looked at in similar light—hazy but precise, originating from a brutal, unnecessary death and lingering in its aftermath. The smoke that refuses to depart, instead affecting everything we call communication—electric, an energy that moves through a conductive path, a charge that can contain immense power, a current that takes with it resistance and voltage;

the imbalance of electrons. If the cloud refers to a whole world of millions of hard-drives, servers, routers, fiber-optic cables, and their networks, Topsy's smoke refers to the silence before death, modernity's obsession with communication and infrastructural thinking, the electric chair and state-sanctioned death, copper's conduction, labouring eternally for capital, incessant entertainment and the impossibility to remain alive and ungoverned. Topsy's presence is persistent, prolonged, and lasting. If in life she refused 'good behaviour', in death she finds electric resistance. Her return to affect sociability is against despair, and acts like a silent weapon to be deployed against ideology that eats itself in order to survive.

Snail's Pace

In 2023, the North-East monsoon saw the super-cyclone Michaung enter the city of Chennai with extreme force. When it was 500km from the city, it was travelling at the speed of around 17km/h. Contrary to expectations, by the time it was 150km away, winds were recorded at 80km/h dumping a very large quantity of rain onto the city. Public holidays were declared and the city received more than 250mm of rain. Stagnant waters refused to drain into the sea or into the over-burdened drainage network that lined the infra-city. There was a city-wide electricity outage for the period of the cyclone.

In this time, those people with homes were trapped indoors. Others were asked to seek shelter in community halls, marriage halls, temples, mosques, and churches. Distracted by tending to broken windows, leaking walls, overflowing drains or backup from toilets, people did not notice that a silent and deadly process was underway in different parts of the city. So when the rains stopped and the waters receded, when the doors and windows of houses and buildings reopened, hundreds of Giant African Snails lining the outer walls, crops, gardens and the ground were finally visible.

The Giant African Snail is a very hardy organism and had travelled from Mauritius to India by way of an English malacologist, William Benson in 1847. He brought two specimens–a single pair–to Kolkata and had gifted them to his neighbour David Barclay. Barclay released them in his garden in Chowringhee before he left for England. Little did Benson or Barclay know that the hermaphroditic animal could through a single copulation carry the innate sufficiency to start an entire colony. While the initial spread of the snail was slow, it gradually extended its reign throughout the then-subcontinent, into farmlands, gardens, and cities evenly right across India to Bangladesh, Nepal, Lanka, and Pakistan.

Since its introduction into the Indian subcontinent and other places in the early periods of global commerce and colonization, the Giant African Snail is now a great threat to farmland and endemic biodiversity with its palette encompassing more than 500 species of crop. The snail also lives in close association with the non-domesticated undergrowth of the city, sometimes living alongside rats, in sewers, in wet areas where there is heavy moisture. For this reason, it often carries a strain of meningitis that is fatal to humans.

The snail is drawn to calcium, a vital mineral that helps grow and harden their shells. They eat bones, eggshells, natural chalk, their own shells too (as soon as the snail hatches from its egg, it turns around and devours its own shell for an initial energy boost).

Calcium is present in limestone, one of the most important materials of the construction industry. Limestone, and its heated version quicklime, have both been used in construction of buildings for more than 10,000 years because of its strong durable and flexible property of withstanding construction load without breakdown. As a material it allows moisture to evaporate, making it breathable and reducing the risk of dampness and mold infestation; it is the basic raw material used to make cement. Lime is also used in agriculture and is present in high quantities in lakes and wetlands, in the production of asphalt, bricks, metals and some medicines; and in manufacturing of bleaching powder, rubber, soap, soda, and paper. The snail is drawn to calcium for similar reasons it is used in construction–fluid regulation, bone strength, shell formation, growth and hardening, cell-wall functioning, and muscle contraction.

So naturally, when not on farm lands, the built city is the snail's source of nourishment. Calcium from the walls of buildings and houses are its calories. The more the calories it consumes, the bigger it gets and the harder its shell becomes. As the snail eats into the walls of houses, it betters and makes more robust its own housing prospects. In the city, human-made and natural cannot be seen as antithetical to each other, and the snail doesn't discriminate between the two either.

It possesses extreme hunger; it never stops eating, eats the crops, eats away at foundations, eats away the earth. This pronounced appetite extends to the rot and decay of the modern urban cities where garbage and refuse pile up; the normal accumulation of material that takes place in parallel to human settlements. The snail, it seems, follows the human everywhere they go. The city and the farmland, and by extension the human, is the fodder that allows the snail to grow and renew itself over and over again. The Giant African Snail has become a slow-moving but demanding shadow of the human that is steadily eating into the hand that feeds it. Neither the city, the farmland nor the human can contend with the ferocious appetite of slime.

A mass extermination drive is currently underway to control its spread. The Kerala government has suggested that a small shallow hole be dug and lined with damp gunny sacks. A layer of raw vegetables should then be placed on that and then covered loosely with wet leaves. This should be left overnight. By the morning all the snails in the immediate surroundings would have slowly made their way to the wet food pot. The next and final step is to pick up the first layer of wet gunny sack along with all its organisms and submerge it into a bucket of either bleach or salt. The Tamil Nadu Government has plainly suggested that people wear plastic bags for gloves and individually pick up as many snails as possible and submerge them in a bucket of salt. They have also suggested that the dead snails be sealed in a plastic cover and put into a public dustbin for incineration.

But the snail resists this extermination time and time again. Many die, but a larger number are born to contend with the culling. Since the snail spends a large portion of its life in dormancy awaiting favourable conditions for its wide proliferation, it disappears the moment the monsoon passes. It rests right beneath the topsoil, waiting, asleep. Pick up a small handful of soil and watch tiny snails and tinier yellow eggs fall through your fingers. This capacity to withstand time has been most profitable to the snail, sleeping and waking up as years pass, retreating and projecting itself, always hanging in the near shadows and waiting to return.

The organism does not *turn to* look at the human, nor does it act in malice. But it does not feign innocence either. The snail, forcibly and accidentally displaced, acts on its own desire to build a more favourable surrounding for itself; to act as its own sentinel. It *must* return every time it can so that its own struggles are not in vain, so that its own surroundings can be changed according to its own desires. If the snail's movement from Mauritius to different parts of the world is a record of history, then the snail is here to act upon that civilizational weight. If the snail was mute witness to war, trade, enslavement, commerce, it is anything but silent now. It secretes a loud incantation of action towards the future where the human and the animal must co-reside if both are to survive. In that sense, the snail is here to reset history, eat away the past and crawl into the future through digestion.

The Giant African Snail leaves glistening tracks as it moves. Though understudied, these slime trails are known to be used as communication pathways between the snails. These paths are records of their version of location, time, and pace. The snail's pace is not merely reduced speed. It is the accident of movement, the incessant need to return and create favorable environments against all odds (including large scale culling). It is the ability to function sufficiently *in spite* of the largest predator: Capital. The snail's pace is hardly a negative connotation. It is instead the ability to stay alive and eat its way through the often incomprehensible appetite of global commerce. The snail, at its own pace, consumes the decay of capital and the delight of capital in the same way; the snail has declared that they both taste the same. The snail's pace is also the digestive pace, the speed of breaking down of food, of the city, of trade, of construction, and of constant growth and the speed of interminable return.

Popping Soap Bubbles, Burning Ashoka Leaves

The stalk of a banana tree becomes the thumping of elephant feet. Unrelated objects create unrelated sounds in highly controlled studios; the more the everydayness of the object, the more its sonic potential. It is in this precision that we find the theatrical, saturated and highly concentrated relationship between the object, its sound, its image and how we relate to it.

In foley, sound is hardly fleeting; sound is not dispersed like in open skies; sound is not explicable. Instead, sound and its meaning are transformed; its perception carefully altered; our ears carefully manipulated to hear another unrelated story waiting to be told. Foley is the phantom in action, a phantom that exists as a precisely attuned craft.

Sounds depart from origin, sounds defy origin. Imagine placing some salt on a dry leaf, pouring some water over it and then holding a lit match to it. The hissing of accelerated evaporation of salt against the slow-burning leaf is sonically similar to a heated iron against a damp cloth, and is also similar to the sound of smoke emanating from Topsy's feet post-electrocution. The squelching of wet paper against a hard surface is the sound of the snail crawling against wet soil. Blowing bubbles through a straw in a glass of water is similar to the sound of the water that sloshes about inside an enclosed coconut. Listen to the popping of soap bubbles, or a dry mud pot quickly soaking up water on first contact as a way to hear the movement of electricity across cables.

Foley creates meaning, new ways of relating, hearing and seeing. First an object is made to produce a sound. That sound is deliberately, precisely and forcefully relocated as originating from another object. It is a play between objects, with sound as the illegitimate oscillating offspring that jumps between objects searching for a home. In that sense foley is not only sound, foley is an event that comprises of objects (parent and foster), the elaborate methods to create sounds from these objects, the people involved in making the sound and choosing the objects, and the new meanings that result from this.

Foley is phantom insofar as we don't see its eventful nature but instead only witness one part of it—the sound and its foster home. The act of listening is a rational enterprise: we see or imagine an object, we hear a sound, and we connect the two. Sound is formally identifiable here. This identifiability lies at the core of the rationale of listening. Imagine breaking open the causal relationship between sonic identification, listening and perception. Imagine witnessing a phantom event instead, where we are carefully unveiling a profound mystery. We see an origin object, we witness the sound it creates and we see it being forced to identify with a completely unrelated object, creating a new causal relationship, a new rational. Foley here is a revealing phenomena—we witness an origin, the sound and its foster home.

Looking for the location of return requires a sense of sound as much as it does a heightened sense of sight. Imagine having to hear the snail first before looking for it. Imagine having to listen to the hiss of burning flesh, or the friction of ropes rubbing into each other before locating Topsy's smoke. Or, imagine hearing the sound of nails being hammered before thinking of the cosmos. Or, imagine the white noise of a mic system signalling a spectral broadcast.

Fantasising sonically goes hand in hand with attempting acute vision. It requires an 'ear to the ground' and 'a fly on the wall' perspective. It requires a sonic invention of a moving physiology of a thousand parts, each expanding and contracting in rhythm at times and haphazardly at others.

The human and the earth become sites for fiction-vision, fiction-listening, tracing-making, object-maneuvering, receiving and transmitting like a radio. All sounds have their object, but can be separated from it. Sounds travel, they appear and then they disappear, they linger in the otolith. Trace speaks of location and a sonic trace exceeds that location. To that end, Topsy, the snails, and the coconut exist in the lingering of their sounds, and are exemplified there. To locate this sonic trace is key to envisioning and encountering their existence.

Incidental sounds—sounds that remain in the background to construct atmospheres, enhance vision—contain within them a surround-ability, but somehow always come to the fore to solidify vision. The envelopment of a sheet of fog, the soft clicking of snails breaking down a city, deliberately and steadily embracing auditory canals, calm atmospheric mid-sea waves carrying a moderately heavy object, the airy ether that circulates wind and myth. Though it is easy to attribute ambivalence and vagueness to ambience since it is everywhere and nowhere, always and never, ambient sounds can disfigure vision precisely. Background scores can be philosophical stepping stones. Inescapable tiny auditory experiences can remind us of situations that have already been written out of history, magnifying the slightest flicker. Incidental sounds render into perception the inaudible, imperceptible, the unattended, the incidental, the unwritten. The intangible smoke, the incantational nail, the slowly growing snail's shell.

Imagine all these sounds magnified such that an entire array of auditory information suddenly explodes inside the ear, travels through the mind, and then defiantly exits through the eyes to see infinitely. If sound is a vessel that helps locate, envision, and create meaning, the cosmic mass that contains elaborate stories sucks them in and relays them across time and infrastructure.

Foley and the cosmic mass are parallel collisions, escaping original cause, or original traumas to return invigorated, with renewed purpose to affect society in ways yet to be discovered. To see this mass requires careful attention; their inter-relational connections are often fragile, tenuous, needing to be brought to life with each telling of a tale. Covert organisms navigate points of connectivity on secretive routes. Transparent and opaque, transparent and opaque. Carcasses, ideas, bits of information, machines, not-machines, a whole lot of things, smoke, snails waiting patiently to resume nibbling away at the city, insufficient incantations, extremely potent cosmic drawings, a quagmire of incomprehensible spools, moving in all directions, simultaneously, all waiting for the right time to return.

ONE MATERIALIST LITANY ON MATHS FOR THE LABOURING ARTIST

THE THIN OBJ
BOLZANO, HERRON, GROTHENDIECK
AGAINST FLOATS: INTEGERS, TEETH, AND THE DEBT CONDITION
DELTA NOTATION

sidony o'neal

THE THIN OBJ

If asked to motivate the study of the practice of mathematics for a mostly non-mathematics witnessing audience, we may first elect to find and speak with a practicing mathematician. Immediately we are faced with an impressive, even maddening, range—thousands of languages, shorthands, and expanding literatures written and spoken not unanimously, but in layered, toggleable tracks. An unruly DAW of maths and its labouring couriers, individuals, collectives, and institutions all blinking and shrieking and ducking and modulating in the so-called arcane and seductive ornamental literature that has been innovating and churning a nuanced record of human experience over many thousands of years.

One thought we recently shared—that mathematics is an undiscerning consequence of human existence—perhaps mathematics acts even as an angle of repose for human cognition. That is, given the condition of existence, we may take that humans can and will make and practice mathematics until we slope, until we can't, until we drop.

Unsatisfied with the answer of the mathematician, we may go to a stalker. Here is a purveyor of abstractions searching for a place to rest upon the arcane itself. Hardly the *maker* of maths, rather here we are after the *philosopher* of mathematics.

Wait—what or whom is meant by *arcane*? Do we make a distinction between "useful" and "non-useful" (making knowledge that is unprofitable within our lifetimes) mathematicians or the mathematics they produce and research? Is there a matrix in the special linear group SL(m) that will resolve the Kantian purposive versus purposeful dyad? Arcane means to say it with your chest, to seal it with your chest. To progress the literature and shutter the library. Lock it all away. Then, there is the major arcana—of infamous mathematicians, of classical math objects, of celebrated proofs, lemmas, and theorems. There is always, of course, more material lurking than this. We have all heard the tale of Ramanujan who labours a gift even on his deathbed, a new kind of number, through Hardy's final greeting which quite literally fails to read the room. There is also the one about the self-silencing mathematician who returned the medal and all of the award money. Have you heard the one about getting it wrong every time, praying for resolution, dying, and then someone getting it right 2000 years later? Or the one about funding trends and "publish or perish?" We will not dwell here now, but in answering our questions we may also familiarise ourselves with the deep and loose handshake between social and spiritual applications of mathematics in popular culture and representation in media.

Alongside and in agreement with Fernando Zalamea's warning from Tymockzo—that the philosophers of mathematics should allow themselves the themes beyond foundations—we have some of our own bridges to construct and theorems to develop.[1] And for a moment our project becomes more or less readable, not unlike poetry, to those familiar with the *lingua vulgaris.* We quickly turn to explore an instance of foundational critique in the philosophy of mathematics that may move on its own accord, a welcome

1 Zalamea, Fernando *Synthetic Philosophy of Contemporary Mathematics* (Sequence, 2009)

ambulance of method and lore, motivated toward aesthetic thought and practice:

First we recall that the "bad company" problem was recently generalised to a classical tension in the philosophy of mathematics. This tension haunts any principle of mathematics which "accepts abstraction on just about any equivalence relation." That is, any principle that allows not an excess of meaning, but instead calls forth a seemingly boundless sequence of (often unbelievable) objects that satisfy similar tastes. As we only require a high-level read of the situation, suffice to say that "bad company" here might be seen as a consequence of "bad" rules and boundaries and we should resist the urge to place bad in the familiar dyad. Bad company often arrives with more bad company in tow and this is a property we aim to discover more about.

Now if such a sequence of objects exists, we may spot it at any succession of points or from any collection of its entries. We may walk in on any exhibition or bounding of the *event*. If the sequence of objects has a limit, we may often be unaware from a single encounter or even from our intuition alone. The recurrence relation, or "formula" if we want to be crass, that generates such a sequence is familiar to the practitioner—the researcher—who may at times recall the notation in place of the sequence itself.[2]

Even as these objects might structure a seemingly non-intuitive environment—consider the quaternion vector space or the punctured plane or an exhibition of new works using materials typical to a given artistic practice—these objects often remain somewhat familiar to their generator. We will call such a sequence of objects

what we've got (or thought) when we're giving our givens.

Returning to bad company, we decide that abstraction cannot be accepted on all equivalence relations—that the principles governing Abstraction are in fact being studied and named and thus bounded and ruled upon wherever an object is found.

Note that the existence of such "unacceptable" *abstraction principles* rests on answering questions about the conditions under which mathematical abstraction, or even "metaphysical explanation" itself

2 I have written elsewhere and again on the powerful and sometimes dangerous fungibility that can be invoked through equation. Even as I adore robust operators, like all ornaments, I must also be aware of their deepest motivations and the desires that brought them into use.

might be acceptable. We exceed the bounds of some disciplining function every time we go searching for the philosophy of mathematics and return furnished with aesthetic meaning.

Let G_0 be the principle that GARDEN ~ GRAVE \iff GRAVE ~ GARDEN.

Then this is a principle which permits such storied company. Bodies and plant matter proliferate whether marked or otherwise acknowledged.

The exuberance of bad company—which for our purposes might be read as the riffy raffy sequence that grows from abstraction *out of place*—is precisely at home in the world of the artist concerned with building a so-called body of work. In fact, our abstraction principles must precisely be those that invite such horrible company, not simply one or nineteen guests, but resound in an unbounded sequence or an unordered array. We search for bad company in the archive of our favourite labouring artists. We may declare, *Ah! This sequence is based on such and such recurrence relation—here in fact may be an abstraction principle at play. And further still—here. This one is truly unbounded. There are still works to come.*

Let the abstraction principles that produce bad company be normalised. That is, consider the clusters of objects produced from such abstraction principles to be ontologically simple. No matter how *bad*, how dense with outputs the practice that produced them, the familiarity of the objects to our abstractionist renders the objects ontologically lightweight in the larger matrix of industry, practitioner, etc. This familiarity grants our cluster an ethereality that Linnebo would refer to as "thinness". We will find that even the objects our cluster contains can inherit this property.

Ethereality is a property we will allow to mark objects that qualify as the *baddest* company. Their principles allow for seemingly endless iteration and exploration on the tiniest of gestures, and yet an experience of these objects immediately refers us to the origin of their relation. We understand that they are amazing together, they know each other. Let this intuition toward ethereality guide our assessment of objects forged not only in mathematical abstraction but in the clumpy complexity of Earth materials and even human relation. For instance, if we want to understand how an art object might be both ethereal yet made using kilos of laser-sintered alloy, or how an object might be simply called a cloud or a nation and require genocide and ecocide on unimaginable orders to exist, we can look to the wide band of affordances permitted by scaling ethereality (ontological lightness, normalisation through familiarity) as a property on bad company.[3]

From here, we might find more intuition for the so-called "thin object."

A smug gossamer among the denser math objects, the TO was introduced to discipline and disappear the kind of abstraction principles (here we loosely mean *rules for making objects*) that inherit or invite the "bad company." This is an object that, if it exists in an absolute sense, is so metaphysically light that it requires very little resource for being, does not "make a substantial demand on the world."

We note that an object may also be called thin in a relative sense, nestled among less parseable objects. After motivating mereological sums, objects result as thin even as their components may enjoy any relative thickness. We can understand from this that the necessary condition is existence

just in case = if and only if

Mereologically speaking, our intuition may also confirm for us that thinner (and relatively thinner) objects proliferate—their generating concepts may be cheaper, ubiquitous, and more or less easy to truss up on a whim.

Linnebo follows Kant in situating metaphysical analyticity as the truly linguistic quest of math philosophy—that is, from the highest heights, the question of whether or not mathematics (as a child of logic is itself) is internally coherent and thus ontologically "generous." The project is determined to depart from a fraught lineage of thinking from the likes of Frege, Hilbert, and Cantor— leading us through a list of concerns within modernist mathematics that unsurprisingly resonate with those in the modernisms of art history. Linnebo's labour over the use of *criterion* and the insistence that we cannot truly approach knowledge of abstract objects leads him to wonder if thin objects "might help with the worry about ontological extravagance"—if it can be shown that mathematical objects

3 An abuse of group theory here might allow us more range through the designation of an ethereality as an "action" on bad company. A thought I leave for myself to work out elsewhere.

demand very little of the world, then "the easier it be to know that the demand is satisfied." We put a pin in the feeling that this thought conjures, coursing through a body which itself is told it demands much of the world to exist, and yet only suspects the inverse is true. From here, Linnebo invites us to consider a rather terse binary claim inherited from Frege. We are then given the direction that "a proponent of thin objects is interested in "sufficiency statements"[4] wherein the statement on the right-hand side of the sufficiency operator has ontological commitments that exceed those of the statement on the left-hand side." For our purposes, as we are *could-be* proponents who often feel constrained by left-to-right constructions, we may instead imagine one strict hexagonal coffin set into correlation with another strict hexagonal coffin. The coffin on the left is empty, printed from a shiny white porous plastic and vapour-smoothed. The coffin on the right is similarly printed and then surfaced in *papier maché*—the adhesive is made from the ashes of the deceased.

If we allow that abstraction is also a "worldly" and "asymmetric matter," where it is in fact possible to generate moderately thick objects from even thicker objects still, then we understand an emphasis on *robust operators*. Operators that must be at least as strong and perhaps twice as secreted as the standalone works/objects/workers themselves.[5]

Linnebo invokes *minimalism* (a useful overload)[6] as a kind of "philosophical magic" that allows one to "conjure up more out of less." We actually find this area rich and well-studied with the inheritance of materials and approach in artistic practice, a refreshing glass of *eau pétillant* for our labouring artists.[7] Linnebo claims that our *minimalist* (*mineralist*) *abstractionist* will answer charges of extravagance by pointing out that after all, in the case of pure abstract objects, this minimalist has only risen to touch—and importantly not exceed—the demands that have already been set by the construction, the inheritance of bounded materials, on the left side.[8]

Further notes on aesthetic abstraction and computational objects beg to fall out of our inquiry, which will take time to elaborate elsewhere—for now, we allow that a computer is a math object first, and

4 The reader is invited to tailgate in the world of necessity and sufficiency statements.

5 Jaleh Mansoor's work on Picabia and "real abstraction" draws us to a much needed relation of stealth, market/capitalism and abstraction throughout histories of modern and contemporary art. In the other direction, insights across Marxist real abstraction miter quite well with some stale perceptions of mathematics as monk work carried out beyond social, economic, and political reach.

thus a thin object may also be a computational object. Then such an object might demand the least amount of heat or land for its retrieval. Any such object may even beg one regime after another to be left alone, to exist as a whole number. It may attempt to live as the reflexive mark it never was. Observe that any environment structured by such objects may call itself "flat-land" even as it allows tangents, sprouts normals, and inherits geometry—permitting unthinkable congruences and excesses sometimes thickened against the very principles that called them into being. Such an environment quickly and selfishly disinherits the labour of recounting its own birth!

The givens continue. Assume that ML models can emotionally co-regulate, and that computers (as machines that process signs) are thicc math objects first.

Endnote: Consider that a few months ago I ran with wintry eyes to the data type that baby me loved the most. I went asking for word-key ~ word-value[9] and instead got the following recurrence relation: Mo ya tué plin moun èk mô lafron; si mo kònyé twa, mo fann tô caboche dan piti mòrso.

I have already killed many people with my headbutt; if I knock into you, I'll shatter your head into thin objects.

BOLZANO, HERRON, GROTHENDIEK

Tu, si quae nosti rectius istis, Candadus imperti! Si non: his utere mecum.

I am, for a moment, embarrassed and refusing to visit a feeling that radiates from the introduction to the 678th and final page in Steve

6 Alma Steingard's *Axiomatics* (2023) develops a framework for post-WW2 US mathematics as "high modernism," wherein we are invited to agree that "in mathematics, the modernist commitment to abstraction reached its logical conclusion."

7 Here I remember Feirrera Gullar's non-object, the Calder dupes, and critical (even trolling) continuations of European modernist intuitions therein. There also continues to be a great deal of projection of the terse alchemy of a meaningful minimalism into this space—from Italian Arte Povera through to its post-national inheritors/interpretations IRL and ONL today.

8 The reader/witness is left to wonder how a number or a set may be considered "pure" in any context as both are wielded often in the name of violent statecraft and ever-increasing restrictions on bodily autonomy. In maths...

9 Here we can see "~" and say "related to" without any undue anxiety over notation. The dictionary referenced is the most comprehensive extant dictionary of the Kouri Vini/Louisiana Creole language. I can't remember what word I was looking up.

Russ's dear writing on the work of early 19th century mathematician, philosopher, theologian, and priest Bernard Bolzano.

Instead, I remember a time a new friend came to an early exhibition, a "methods" show,[10] and guessing at some (on everything—the thoughts had never crossed my mind!) geographical routes into a terse series of processual drawings, gridded abstractions of soft marks that deftly hid most of their abstraction principles, the new friend ceased with a soft jape, a confident surrender, *Ah! You must not be as traumatised as the rest of us!*

Swiping back to Bolzano a bit more encouraged—I'm reminded that not all objects and intuitions which may appear to us need to have a form. Thus all concepts are *a priori,* as there is no way to defend the binary imposed by the assumption that (especially within mathematics) empiricism is possible. If mathematics is a "subject" where (spiritual, affective) intuition, and often facility, sharply decline with lack of experience, then it is not unlike other subjects which may be called non-scientific but similarly rely on building and maintaining intuition for sustained research in abstractions.

One morning walking in Paris, I am making jumpy notes aloud about the normalisation of a very old ornament that I find recurring in the architecture of a few major European cities (I highlight the exception in our recent trip to Rotterdam with it's history of razing and modernist rebuild) and its relationship to conversations on ornament in postcolonial aesthetics in the Americas. I think of my mother, I think of the ornamental objects that came from thrift stores and markets that lived with me in my childhood. I realise that I am obsessed with Russ's picture of a dying Bohemia and the material and physical conditions that can be used to contextualise the work of a young Bolzano. I am obsessing because of something that has been obviously, annoyingly, stuck in my teeth since the soft jape.

Where there is trauma, abjection, difference—whence abstraction? A construction that feels too simple because it is, but I am in no position and truly without the time to ask better questions. The reason I consider it again here—well, Piper Harron's work is what I really want to talk about now.

10 I used to make a categorical distinction between the kinds of shows that present "methods" over "applications" and vv.

Nearly a decade ago Harron's dissertation *The Equidistribution of Lattice Shapes of Rings of Integers of Cubic, Quartic, and Quintic Number Fields: An Artist's Rendering* went viral on math Twitter. Harron's humorous and sharp work asks us to look at the conditions under which mathematical abstraction may be supported or even undertaken at all. As we are led through the proofs and the motivation that make up her research, we are met with unconventional "laysplanations" that describe the vast number theoretical and analytic developments in the world of lattices[11] while keeping in mind the mathematician's experience as a racialized, gendered parent—a person in a body existing in a world of climates and public health crises who's work and occupation is deeply cognitive, internal, and demanding.

Two summers later, I share Harron's work with students (a mix of visual artists, computer scientists, poets, writers, etc) at the New York-based (and often administered online for students based all over the world) School for Poetic Computation. I am co-teaching a class I often nickname "math for artists" and I feel like someday I should also figure out how to support a "contemporary art for mathematicians" class—I want students to hear that as practitioners attempting to find/make/develop abstract thought over a field of ever deepening cognitive abstractions while existing in fleshy vessels requiring shelter from weather on Earth, we are taking up some of the least understood activities in the history of people. The fear of being useless, of adding nothing to the world—a fleshy thought that belies its own capitalist entrails—is always creeping and it is valid. There were many mathematicians, practitioners, PhD candidates from PWIs who shared Harron's dissertation online—shared in experiences of marginalised gender, race, ability, and class within the small circle of researchers in pure mathematics, and notably within the bad boy's club that historically has named itself Arithmetic and is now known as Number Theory—everyone was like *please look at this shit RTFN.* Years later, it was my undergraduate maths advisor who'd done his own graduate work in number theory in the same

11 Not super important for our time here, but lattices may be understood as a specialisation of an algebraic structure, a *math object*, called a partially ordered set. We may visualise a partially ordered set as a collection of objects where "order" or "hierarchy" or other comparative action may be taken upon pairs elements of the collection to a reasonable degree. Observe that the *rhetorical object* "you cannot compare apples and oranges" exists across many languages and cultures, with relevant substitutes for the produce in question, and cheekily rhymes at a high level with our interest in posets.

department as Harron—after hearing my experience of twinned excitement and embarrassment, and that it had taken so long for me to admit a desire to seek qualifications in perhaps aptly-called "pure (lol or historically-unapplied) mathematics," who suggested I take a look at her text.

I was single-minded about my own Twitter use at the time. I had been compulsively compiling and retweeting and posting on mathematicians who were or are African or of African descent and living in diaspora. I was trying to determine what advancements on their doctoral works had been made, what kinds of funding they may or may not have received to continue their research. For many reasons a PhD alone, or even academic validation, is not enough. A mathematician's work begins long before, and if so blessed, will extend well beyond doctoral research and the work of one's doctoral advisors. I read through Harron's work on lattices and being Black and occupied with high order abstraction and anxiously wondered where the work—and where she—might end up in the decades to come.

Bolzano's work was hardly celebrated during his lifetime.

The Prague-born child of Italian immigrants spoke and published in German and lived much of his early life under the complex regime change that ushered The Kingdom of Bohemia into an era of liberalism.

What motivation for a turn to the simplest of abstractions—in this case many routes into proof of the line or the circle—could Bolzano have found? Into which cracks must the volcanic ash of trauma settle in order to look beyond any material application for one's clearest thoughts and instead publish the words "I intend soon... to give detailed criticism of *all the defects of our previous systems of geometry*."

Bolzano's work on what he called "The Three Problems" carried from his intuition that some fundamentals, seemingly untouchable *arche* could be done away with—for him these included the whole of infinitesimal calculus as well as Archimedes's axioms— in order to arrive at a more believable "science of space". Bolzano's writing easily feels prophetic, the kind of proof-making and theorem-motivating that only a theologian could set up and execute. His expositions contain spicy charges, heady invitations to all mathematicians, challenges to those who could claim to scrutinise his work. Bolzano was outspoken about the militarism that he felt had plagued an increasingly expansionist kingdom and eventually he was forbidden from publishing in both mathematics and theology altogether before being exiled.

I fuck with the whole project heavily. Bolzano's ideas range and thicken in places where you might least expect. Elaborating frameworks like *Theory of Science* or *Consequence and Ground* feel like they took a lifetime to develop and yet remain largely inconclusive or not-yet-applicable. His writing is cool, petty, and driven to dazzling illustration in the ways I hope to show up as in-person, IRL. Bolzano's facility with Greek one liners and tight classical maxims directed at his colleagues is expected. His words are inspired and sly in the way they fold up time and faith in the logic of many literatures. Mathematics with all of its constants and variables is, after all, a special form of literature.

But here we are talking trauma, even material. I look again. Bolzano turns to math in adulthood, after priesthood.

Further elaboration of Bolzano's natal family reveals that 10 of his 11 siblings died before reaching adulthood. I count the ways the funny, spicy, observant, and skilful abstractionist rhymes with the characteristics of many people in my own life who live with deep and often recurrent or chronic trauma and/or illness. It is from here that I am convinced that deep abstraction belongs to us.

I want to invite now something like a smallish lemma about our Alexandre Grothendiek, the feverishly prolific child of French anarchists, the anti-imperialist who undoubtedly set a thorough course for contemporary mathematics and then promptly recused himself from the discipline in protest of those atrocities protected by the state as the academy and the community of scholars—the mathematician and philosopher on whom an endless scroll of writing has recently been published across popular, fictional, and scholarly contexts.[12] For our purposes in touch with Bolzano and the question of motivation, we will reconnect to this note from Grothendiek's diaristic set *Récoltes et Semailles*:

"Les violences que j'avais eu moi-même à subir au cours de mon enfance depuis l'age de cinq ans, sans même avoir jamais été désignées comme telles à mon attention d'enfant, vient fini par créer un état de tension chronique, réstee inconsciente et soiguesement contrôlée par une volonté bien trempée..."[13]

12 See Carlos Fonseca's *Coronel Lágrimas* for my favourite of the fictional writing.
13 Récoltes et Semailles, II (1039)

Grothendiek recalls that the violence we endure in childhood can leave us with a restless tension, and a need to "control" the "well kept secret" of anxieties that are apt to resurface into adulthood. He goes on to call "free violence" what someone else might call living with complex posttraumatic stress disorder, or what someone else might use to build an art career. He attempts to clarify—that animosity which is "non fondée ou provoquée par des griefs concrets..."[14] Grothendiek's animosity here, aggressive motivation with an unclear anchor, is both free as in the root cause is so deeply masked, or free as in it can never be clear to us how/why the emotion appears to cost nothing to exist. It may be experienced as thin in that way. However, we may also read that such violence is *free* as in untethered or has no mandate/bosses/is sovereign or and is thus given to range as it pleases, through our experiences in relation to family and into our many generative/creative practices. It should surprise us very little that Category Theory, Grothendiek's most famous mathematical project, is primarily concerned with mapping relation among math objects as though such objects were free. This shift to a method that generalises relation on objects across the many capital t *Theories* in maths is both philosophical and political in the history of maths as it exists through and in the world. Upstaging a modernist set theoretical impulse to segment and enumerate and individualise, Grothendiek's Category Theory seeks to describe relational affordances, fluidity and connectivity among disparate objects. In another context, its primary concern might even be sublimated to an ambitious democratic project—the study of equitable connection of disparate entities (defining useful tools, proofs, and related functions) through the well-defined preservation and respect for difference and transformation. We will end the lemma here.

Under Western forms of capitalism, useless curiosity (non-applicable abstraction) is dirty. It is curiosity out of place.[15]

This is an argument that has been made with several faces, some of which I agree more with than others—again I feel somewhere that the purview of abstraction, and direction-setting for contemporary methods in abstraction, most critically belongs to those who

14 Ibid. (1042)
15 Palmerston, Douglas, et al., basically context determines what is pure and what is not.

live within those matrices of flesh, debt, material, (hyper/in)visibility that make it hardest to turn attention to abstractions which are not readily applicable. Those who are intimate with such violence are also fluent in the kind of abstraction that takes its time. It is an abstraction older than the debt condition or the human relation of enslavement. It is Indigenous abstraction and of course not relegated to the activities of human thought alone. Violence can never truly be free in that it costs nothing to exist, but rather we may observe and critique that when violence roams normalised, an ethereal object scaling in relation to and within our lived experiences, it calls itself free.

Finally, embarrassment has always only been a secondary emotion. Embarrassment is also mistaken grief about the uselessness associated with dominant discourses of abstraction as niche and rare—especially those grown in the context of western hegemonic logics. For many reasons visible and otherwise, one might live with the inheritance of ready utility as a property we must demonstrate and maintain to belong to the labouring and desiring world. Brilliant abstractions and abstractionists no matter the discipline, no matter the epistemological framework, are not sheepish and *uncommon*—rather they are silenced, exiled, and readily killed/killable in a world where everyone's survival depends on our inheritance of, and our continuation and disruption of, logics and practices of abstraction. Can unpacking this intuition lead to a useful question? *Archi possible.*

NO FLOATS: INTEGERS, TEETH, AND THE DEBT CONDITION

The teeth are chattering out of their heads:

A: ATTENTION, ZAHLEN, ZAHNE, ZHOLLEN,
SWOLLEN, GOLDEN, TROPHY, HOLD THEM.

B: DENT, DEBT, DEBIT.

C: "We thus hope to make evident this idea that the true logic is not *a priori* in relation to mathematics but that for logic to exist a mathematics is necessary."[16]

16 Lautman trans Duffy (2011)

When your teeth fall out as a kid you wait eagerly for them to grow back. We will call on this expectation to approach the embodied numeracy intrinsic to the debt condition. You keep track of the teeth. You bury them under your pillow. In some traditions there is money left in the morning. In other contexts the teeth never belonged to you in the first place.

I found a bag of my teeth in my mother's top drawer one year. The tops were permanently stained brown with old blood from my gums. We are not talking about familial debt—the debt you accrue through a unit of relation that has become central to survival (or at least assures a modicum of institutional relief and relational care over an orphaned person/"single" individual) under capitalism—although many relationships to negative numbers could likely be traced to expectations within families chosen or otherwise constructed and maintained.

Deluezean constructions seem equally concerned with/contingent on notions of continuity and—like I ask the tooth fairy's money for my teeth—had I been there, I might have asked for more discrete routes out of Lautman.[17]

In some frameworks, every byte/bite is time. I don't want to have to tell you that time wasn't always money. A brief history of maths in finance would do well to include teeth. At the supermarket and at the bank I am prompted by both humans and computers to round up, to donate the change from my purchase, to transfer one USD or more. *Would you like to make it a (nice, positive) rounder number?*

Yes, ok. Just for tonight. As long as the number is positive (not zero) and rounded, it cannot be preceded by that stylish knife of negation, it cannot be negative. Computers are afraid of (bad) floats, why does their discreteness freak us out?

It's a goofy thing and a mad thing losing your teeth /neq losing your marbles. Having your marbles taken.

17 One nice thing is that Lautman turns the line into a "covering surface" into a torus. After thinking about this for a time I am invited to a conference where it is proven to me that the moduli space of the torus is a thicker object than the moduli space of the triangle (which is also a triangle.)

In the states, the history of "Waterloo Teeth" and other traumatic traditions of trophy teeth-taking—from the bodies of dead and prisoners of war—can show us how teeth (and their absence) have always spoken to the debt condition. We are presented with such warnings all the time. People addicted to substances or with disordered eating behaviours or elders and chronically ill people losing vital nutrients over time—anything that makes it harder for teeth to stay in one's head—become more visibly "integered" over time. Socially we observe something like a zero-mirror across which positive integered and negative integered entities begin to take on affective qualities. In the black. In the red. A debt of gratitude. You owe it to yourself.

One construction of \mathbb{Z}[18] is as follows:

Z is the set of equivalence classes of $N \times N$ where $(a,b) \sim (c,d)$ iff $a+d=c+b$.

An abstract algebraic definition generalises the construction of the integers to localisation (introducing a multiplicative inverse to a ring).

For our purposes, take the numbers you can count by succession—imagine you have no teeth and then slowly teeth begin to grow and might continue sprouting without cease. Create a copy of the boundless teeth and then make a definition for what it means to multiply them by (-1). Ah, don't forget that an *identity element* called zero must exist, assign this to the state of *just gums*. So now we have a boundless ring of teeth and their copy multiplied by (-1), and an identity called zero. We will call this world integers.

After watching a coming-of-age film, I tried to snag out my own very loose tooth out by attaching it to a string (was it fishing line?) which was then attached to a doorknob attached to a bedroom door. It's maybe 1998 outside and easily one of the hottest days of the year. The door moves through the air quickly but I still have to slam a few times to get the tooth out. My body's pain cocktail quickly rushes to meet the need. The debt condition as it relates to feeling cost always requires that you are willing to give something up, even better if it's something that *matters* to you, in exchange.

If you have nothing in mind to exchange, then it means you are unwilling to grow (up).

18 the LaTex macro for the fancy version of the Z Hilbert assigned to the group.

I prompt participants in a meditation on math trauma that I am leading to hold a number in mind and to transform that number freely through a series of invitations. Later I ask them to recount or share a number that feels good to them in the moment. I offer a small negative number—a negation only four digits long. It takes little time to compute, barely creating a stir/heat and thus is costs me next to nothing to offer it to the room. A student quickly remarks how unheard of it is to bring a negative number to the "favourite number" game. Another student offers a number with a decimal and digits out to the hundredths place—a float!

Zahlen, the German word used to invoke the integers, has a few meanings including *to count* and *to pay.*

The credit system in this country profits from the same anxiety that attends the negative float. The same anxiety, induced by waiting to cross over zed, causes me to overnight ship—it asks us to make up a balance or to learn how to live with the action of negation. Negation is not nothing, but it is an action that truly requires the existence of nothing in the first place. All I know is nothing is more important than soothing excoriated gums. Growing back the teeth. Even when I know the condition is temporary I want it to end. I need my teeth to grow back every time. I will fake it if I have to. The numbers on our credit card statements increase (bad, positive) to mark absence in the debtor imaginary. From here, unless you are a corporate proxy, only debit is honest.

A computer is an ontology-producing machine. Recall that no matter how endless they seem, computers are discrete machines and must eventually find a way to bound all floating point numbers aka floats.[19]

Fuzzily speaking, the *Softmax* function is also a kind of normalizer that works (normalises data) for attention. Like the presentation of credit debt, rather than imagining an endless promise of teeth, it rescues us from the uncanny result of a prediction algorithm by spitting out positive ints.[20] A simpler example of the same

19 Single-precision floating-point format accounts for the unthinkable amount of numbers a computer can format using 32 bits of memory. Still this amount of numbers as well the number of places beyond the decimal are bounded.

20 Here I recover the function from a paper presented at NIPS 2017 beautifully titled *Attention is All You Need.* The paper would go on to inspire many directions in the NLP and machine learning research space, ultimately providing the groundwork for a simpler Transformer (one which evades potentially fatal recurrences and the messy composition of functions) and speeding up much of the parsing required for training models. Clocking in at ten pages followed by a few pages of graphics, the paper has become a legacy text in the NLP world.

brand of functions might be something like the ABS function. ABS would make a lovely propaganda machine—it turns negation into progress. It turns travel into direction-agnostic distance and simply reminds you how far in any direction you are from zed. Anyway, what makes *Softmax* a weirdo is that it makes positive integers out of negative possibility. It exists to set fire to the uncanny valley invoked by deciding that something is *this much* likely to never happen.[21] A useless figure to hold unless you are Funes el Memorioso—are we ready to talk about the material impossibilities and necropolitics inherent to hyperscale computing?

If you dream that your teeth are falling out, you might ask the internet and be told that the possibilities for interpretation range widely depending on context and language. If you dream that your float is algebraic and unbounded, you might wake up and call it transcendental. In any case both the body and math object are ill-equipped to hold it all at once, and even in relation, no matter how abstract the object, there will always be debt—land, weather, blood—to normalise.

DELTA NOTATION: THICKER THOUGHTS ON THE ORDER OF DERIVATIVES[22]

/delta

The delta.

The Mississippi River valley.

The incremental shifts that humans imposed on the Atchafalaya until it became the Mississippi.[23]

For the sake of our inquiry, works may follow from the place where algebraic operators grow, from the study of analytic calculus, functions and curves that carry cooling devices through settler-applied mathematics (again, we allow that a computer is a math object first)[24] into industrial ornament, from the meaning of the first tree (in the

21 One interpretation of "negative possibility". The reader is invited to think through a few others.

22 Some of this writing has previously appeared in *M:ASR* a publication associated with the exhibition *ENCHIRIDION: AISLE, SPLINE, RESORT* which took place at the Portland Institute for Contemporary Art (US) in spring 2022.

23 See Mcphee The control of Nature, 1989

grave/garden) we peed underneath, from an apotropaic use of the fleuron (see: rotated floral heart bullet)—renders from ongoing research on post-constructivist methods across art and maths.

If ever I am asked to recover *the state* using smooth integration by parts, I will once and forever refuse. If ever I am asked to recover *race craft* I will point to Ramon Amaro and Murad Khan when I say that Black existence is not analytical, that is its logics are not coherent with the mathematics called "what it means to be Black in a world regulated by the substance of race."[25]

Motivating an interest in the tools of calculus, and especially unpacking the stakes of higher order calculus, requires intensive two-fold mouthing which truly takes more time than we have together currently.

It is sufficient to note that among some mathematicians, following constructivist tendencies, it remains in open contest that the continuity of \mathbb{R}, the assumed boundlessness (in both length and density) of the real number line, the capacity to "glide" or to / delta without skipping is...a reach. It strikes me then to allow the following procedural collapse:

The celebrated proofs of 19th century functional analysis and the calculus of variations depend on rigorous efforts to revise and discretize Indigenous non-European ontologies or meta-ontologies or whatever we are calling the life of our being these days. For Amaro, the calculus of Newton and Leibniz, the calculus of variations is itself the "concretisation of the black individual." First we should decide what may be felt by unpacking the calculus of variations. What topological properties exist for such a sign in the artist's world? What properties of the word-object hold across contexts?

24 Magna pace Dwyer (2023). My position on math, collectivity, and the seeming redundancy of consciousness in mathematics across human time and experience often leaves me unsure about how and what kinds of mathematical thought to say "belongs" to what group. I am interested in the legislative, economic, and political *applications* of mathematics only insofar as they can tell more about who has been most privileged in engaging, publishing (both popular science communication and internal academic research publishing), and the de facto patenting of mathematical thought. Rather we can speak of a *Settler-applied* mathematics which builds value against western cultural associations with rectilinearity, while ignoring the use of complex and vectored/planar/array-based geometries that have existed and are being made even as a mundane matter of course across many Black and Indigenous and non-western spatial and spiritual systems. Settler-applied mathematics refuses a strategic generalization of the objects of linear algebra beyond the vectored line and into the study of matrices. It is an applied mathematics that secrets knowledge that might otherwise liberate.

25 *Towards Black Individuation and a Calculus of Variations*, in *Navigation Beyond Vision*, Sternberg Press, 2023 (192)

In a wargaming environment there is nearly no space for asylum.

Thus, the refuge I seek in earth/moss world must also be a refuge of approximations, a refuge of *nearness*. A refuge that treats the linear approximations fundamental to calculus as Indigenous and inherited techniques—moving diasporically on the mainlands and beyond—held among other techniques in the sacred study of increment and variation, difference and bo(u)nds.

Synthetic requests ~ synthetic h0sts /rightarrow eternal requests ~ eternal h0sts.

This intuition can take more time to unpack than I have now. In brief, it means taking seriously the feeling that the objects/nums required to make something called *le calcul* are called *real* only when they obey or clearly disobey a communitarian prayer for mostly smooth wholeness. These are objects whose nearness must be described (or constructed where possible) and consented to before proceeding.

This intuition is useful to approach higher order analytic contingencies anyway.

One recurring thought is that the symbolic derivation and estimation of values through gorgeous maths techniques contained in The Calculus of Newton and Leibniz was perfected alongside the extraction of value from human and non-human life and labour in diaspora. How to render a post-object personhood against the equivalence of symbol and 3space proposed by some Platonist and post-Fregean lines of thought? What if we proceed from the axioms of ZFC set theory— logic that allows for colonies, blood quantum, arpents, and hectares? Those principal terrorizing functions and partitioning strategies were studied and built mathematically alongside the creation of the same settler states that now promise, even mandate, forms of *refuge, security, asylum*. Even "dynamic" set theory appears to have been made for lovers of ossified partitions, plantations, and wargames.

In 1930, W.R. Ransom, then President of the Association of Mathematics Teachers in New England, reported that the trisection of angles was a conceptual *bête noire*, which I interpret with and beyond its usage to indicate a racialized *problem for thought*.[26] From here it is possible to revisit the Homestead Act of 1862, itself marked by a tridential procedure for settlers, which clearly mobilized analytic and differential geometry against Indigenous and Afro-Indigenous life and/even as it invited the participation of some US based descendants of American Chattel Slavery. An ideal mesh (that workhorse of geometry processing) free of pinches, redundancy, or nasty "foldovers" inherits and acts as the survey systems of state bureaus of land management. Part of the matter, a lemma, includes chasing back some so-called abstractionist gestures, down the branches, down down down, until you find yourself in the not-always-discrete and not-quite-continuous world of mosses. Never again enticed to mistake tree-climbing for flight.

Maybe you are like me, whispering, *but I am only a math object curious about other math objects.* And then again, sotto, *resort.*

26 cf. Nahum Chandler. Here I am less interested in the use of maths philosophy to build maths pedagogy. I address Ransom's thoughts on "casual" and personal algebras—abstraction—more directly in a forthcoming project.

THE LIGHT OF THE SUN MINUS A FIREFLY
Oscar Santillan

"Get ready, brother, Christ is coming soon, like a dragon in the night!" proclaimed Caribbean evangelist Yiye Ávila during one of his massive "salvation and healing campaigns" in the early 1990s. That night he held the microphone from a makeshift pulpit in the middle of a half-filled soccer stadium. The grass was yellowish, hundreds of crickets and all kinds of bugs were being spurred on by the stadium's immense floodlights; insects were falling, scorched, raining down like biological meteorites. The poor of Latin America were leaving the Catholic Church for an even more orthodox—albeit more ingenious—Evangelical Church. Proof of this wit, and of this under-appreciated link between evangelical Christianity and science fiction, is a story told by the preacher that yellowish night: "Some Russian scientists sent a spacecraft to far away places, beyond Jupiter, and it was moving away and away from Earth. The spacecraft came to a huge rainbow that sucked it in; they say it passed into a fourth dimension, and suddenly it came to a great golden city, into the heavens. (Praise Him, He who lives!)." There it took a hundred pictures that have been hidden by these atheistic communists; and, there is also an audio recording in which you can clearly hear a divine voice warning humans: stop traveling around the universe, this is my abode, your's is Earth!

For young evangelicals in Latin America, Yiye Ávila was the closest thing they had to Carl Sagan. No one in the crowd seemed to spot the many incongruities in this story. For example, that audio recording was impossible since sound is not transmitted in the vacuum of outer space, which is why space probes do not have so-called voice recorders recording nothingness. The preacher's brash arrogance also concealed something more noble, an involuntary stimulus to the imagination. Though distorted, this is the introduction to astronomy that thousands of Latin Americans have received.

A few years earlier, another Caribbean man had been the first Latin American astronaut and also the first Afro-descendant astronaut to participate in a space mission. Cuban Arnaldo Tamayo left the Earth's atmosphere aboard the Soviet Soyuz 38 spacecraft on a cloudy day in 1980. Over the course of a week in orbit, Tamayo circled the planet 124 times while conducting a series of scientific experiments, including some very intriguing ones:

- Study of the growth of sugar crystals under micro-gravity conditions.
- Changes in psychomotor coordination of left and right hands.
- Study of space adaptation syndrome (better known as "space sickness").

It took Tamayo just over an hour to circle the Earth each time. As he did so, he held a grain of sugar between his fingers, tracing a circle in infinite space.

The notion of an immeasurable distant space beyond the planetary atmosphere is in profound conflict with the information perceived by the senses. Thus, human eyes are unable to indicate that a small point of light at night is actually a star larger than the Sun; and, the skin can feel the heat produced by photons of light from the Sun as they collide with the epidermis, but the skin is unable to feel the photons of light from Alpha Centauri, the nearest star to the Solar System. It is also the case that the spectrum of light that human eyes perceive is very limited. The ultraviolet light spectrum and the infrared spectrum exceed this perceptible range, while many insects—with extremely sophisticated eyes—are able to experience this expanded chromatic reality.

What is the distance between the stars and the billions of eyes that observe them from Earth? For thousands of years an insistent hypothesis was that the sky was a dome—a rigid one—that enveloped the planet. In it all the stars and astronomical phenomena took place: there was room for the entire universe. Sumerians, Egyptians, and later Hebrews, subscribed to this notion. This ancient conjecture still lives on in language: "celestial vault" or "firmament" (suggesting something firm) are nouns that today are used interchangeably to refer to the sky, which is another confusing noun describing the arbitrary sum of the atmosphere plus the immensity of outer space. Thus, for thousands of years it was thought that the universe was only the size of the Earth's atmosphere; then it was thought to be the size of the Solar System; then it was thought to be the size of the galaxy that contains it, the Milky Way; then it was learned that this galaxy is only one among 100 trillion other galaxies that constitute the universe; and, then it was intuited that the pluriverse would be possible: big bangs generating innumerable universes outside the space-time experienced by humans.

Tamayo returned to Earth as "the Yuri Gagarin of the Caribbean". Both astronauts shared the same proletarian origins and the same smile, which produced sympathy even to "the enemies of revolution". The Cuban dictator, who had sponsored this space trip, welcomed Tamayo with a remarkable event that included one of his long speeches. A mixture of pompous tirade and Marxist self-help manual, the dictator's speech included a surprising detail, arbitrary perhaps since his gesture did not seem to tie in directly with the rest of the speech–he read a haiku. Cuban television broadcast the whole thing live on its single channel. The beautiful baritone the dictator possessed, and the minimal but careful reflective pauses with which he delivered this reading, allowed for an accurate appreciation of those three lines written two centuries earlier by Oshima Ryota:

Pursued,
the firefly,
hides in a moonbeam.

**

The astronomer Francisco Díaz Covarrubias is one of the driving forces behind the construction of the observatory that in 1863 was located at the Chapultepec Castle in Mexico City, although perhaps the discrete trail of recognition that has followed him is due to an event that took place in 1874 when, along with four other Mexican scientists, he embarked from the port of Veracruz bound for Japan.

After two centuries of almost absolute isolation, just a few years before the arrival of these five visitors, Japan had been forced to reopen its doors to the West. European capitalism and enlightenment, packaged in the form of political liberalism, were then continuing their campaign of global expansion, and modern science was consolidating itself as the new "universal" ideology that would "liberate" people from "primitive superstitions". In its unique ability to explain the world in the imposing language of mathematics, by the 19th century, science had been able to narrate the world as the unfolding of natural laws. Since then, mathematics not only dominates the "hard" sciences but also, in the form of statistics, plays a central role even in biology or anthropology; there, statistics process data collected empirically or in field studies, giving them the appearance of accuracy.

A photograph of the five members of the Mexican astronomical expedition has survived from that trip to Japan. Posing for the camera, they look serene, dressed in European-style costumes, just as European as the ideas they carry with them.

The progress that they cherish is the West's, which was the only future they could imagine and the only past they knew and could aspire to belong to. Three hundred years of colonization had obscured a past that, in the 19th century, had not yet been returned to the present. The new Criollo elites who imposed their political power to end colonisation were bipolar, their flamboyant nations recognised themselves in the mirror of European modernity, while at the same time aspirational, urban Criollos did their best to hide any indigenous or mulatto vestiges. The image reflected by the mirror was unstable, mutable, refusing to unify the reflection into the consistent identity that they had hoped for. But, for the Japanese the distinction between national identity and modernity was perhaps clearer. The Japanese reformers—who led the rapid industrialisation of their country from the second half of the 19th century onwards—described it in their motto "Japanese spirit, Western technology". Therefore, they turned that tense duality into an assertive vision.

A thousand years earlier, Mesoamerica had had epic observatories, astronomical treatises, and hundreds or thousands of experts who studied and computed the movement of the stars, though all this was unknown to the five men posing in that 1874 picture; even in that year the ancient Maya civilization was barely a shadow buried beneath immense jungles. Precisely a thousand years ago, while the civilisations of Mesoamerica were experiencing their own "Renaissance", Europe was dominated by the most rampant and fearsome religious intolerance, a land overrun by plagues that spread through the filthy waters of many of its cities, and its mathematicians were ignorant of the number zero. It was a place far removed from the technological splendor of China that had invented gunpowder, the compass, the printing press (yes, the printing press with movable type was invented 600 years before Gutenberg), and the most advanced naval engineering. Europe was also a stranger to the scientific innovations that the Arab world was promoting on the basis of the Greek classics that they had saved from the Christian bonfires, and the mathematics they had learned from India and Babylon (equations without which it would never have been possible to reach the Moon or Mars). Unexpectedly, the self-proclaimed "old" continent was to become the world's greatest cultural and scientific hub from the 16th century onwards. The dogmas of medieval Catholicism were cracking as humanism and Protestantism emerged as powerful ideological alternatives. Silver from the Potosi mines—which doubled the circulating capital of European kings—would pay the bills for these renewed ambitions, which included the creation of great engineering projects, new academic institutions, and patronage for scientific advancement. Already by the 19th century theology was a domain separate from science, just as—as the liberals aspired—the church would be separated from the state.

One of these convinced liberals was the astronomer Díaz Covarrubias, for whom it was clear that there was a link between his scientific work and the social "progress" of his country. The expedition to Japan that he and his colleagues were about to embark on that autumn of 1874 was scientific, but for him it was also essential that it should "inspire modern ideas in my compatriots". The expedition set off by train from Mexico City to Veracruz, from where they were to embark on a full orbit around the planet. Covarrubias is 41 years old, with a marked agility and somewhat aristocratic manners. That morning in the harbor he seems rather agitated; he approaches the captain of the ship to insist that he does not postpone departure, continually checks his notes, at times converses with his colleagues, and at times finds himself engaged in conversations that do not really matter to him. He carries with him a shipment of large wooden crates containing the best astronomical observation equipment his country has to offer. There is an unwanted possibility that the five scientists and their cargo will not make it on time to reach their final destination. The route to Japan was long and at that time it was not possible to set sail from the Mexican Pacific because there were no reliable routes linking the Mexican capital with the west coast of the country. On those downed routes, observation equipment would have been ruined. From Veracruz, on the Atlantic coast, the route continues to Havana, from there to New York, across the United States to San Francisco; they sail for weeks until they anchor in the port of Yokohama, Japan. Covarrubias is concerned that the execution of the itinerary may not be achieved with the haste they are carrying, and also that the precious equipment should not be out of caliber or damaged. He personally supervises the handling and placement of the cargo at each station and port.

Another member of this scientific expedition was a journalist and engineer, Francisco Bulnes; both were to write two books containing their memories of this long voyage. Bulnes wrote a memoir with a latent literary sensibility which he entitled *Once Mil Leguas* (Eleven Thousand Leagues), while Díaz Covarrubias' tone was more scientific and historical in his *Viaje de la Comisión Astronómica Mexicana al Japón* (Travel of the Mexican Astronomical Commission to Japan). Both books are precious documents of the very first contacts between two universes separated by the immensity of the Pacific Ocean. Finally, at midnight on 8 November 1874, the ship the scientists were on dropped its anchors. Covarrubias looked out of the air vent of his chamber and immediately noted down these first images, "the clouds had almost completely disappeared, and in the dim light of the stars I could only see the Asians in the boats as shadows for the time being. At a greater distance and at a certain height above the sea, the black masses of the hills surrounding Yokohama on the land side could be distinguished, limiting the horizon to the distant mountains on which the dominant peak of Fusiyama stands out".

The next day, preparations start already, and in a month's time, on December 9th, a rare astronomical event will take place: Venus will align perfectly between the Earth and the Sun. That is the mission of the Mexican astronomical expedition: to observe and time the duration of this event, which is known as the "Transit of Venus".

In modern history, this is the first international expedition in the world led by Latin American scientists.

***

All distances within the Solar System are measured with a unit known as the Astronomical Unit (AU). Since the measurements of length used on the planet, such as metres, inches, or leagues, are based on small human scales—another reference point—an extraterrestrial starting point was needed to establish a unit that better represents these immense distances. Thus, 1 AU is equivalent to the distance between the Earth and the Sun.

For at least two thousand years attempts had been made to establish this distance, Greek and Chinese astronomers being among the first to have made this calculation, though with somewhat imprecise results. It is likely that various attempts were made in other places and times; for example, one civilisation whose mathematics, geometry, and astronomical records would have allowed it is the Maya; but, the Dresden Codex—a sophisticated astronomical treatise—does not record this information. This codex is only one of four Mayan books that survived the fire of the dreaded "campaigns of extirpation of idolatries" that the Catholic Church carried out in the Americas during the 16th and 17th centuries in order to exterminate the knowledge and worldviews of the continent. Written around the 10th century, its pages contain precise mathematical tables describing the movement of the Sun, Moon, Venus, and Mars. This careful recording of the movement of the stars has a clear purpose in the Mesoamerican cosmovision. The data is processed and incorporated into calendars that assemble the diverse oscillations of the stars with the various rhythms of life. The Mayan astronomers, who were an intellectual class of priests, had perfected a series of multiple calendars that mesh with each other: a solar calendar of 365 days, a lunar calendar of 260 days, another of 584 days, in accordance with Venus, and others; all geared with absolute precision in a great cycle of 5200 years: the great cycle is an immense lake in which the streams of different rivers of time converge. I like a cenote of liquid obsidian in which the future and the past are submerged together. Mayan eternity is thus not linear but cyclical, renewed every 5200 years.

The Gregorian calendar used in the West today—a 365 day solar calendar identical to its Mayan counterpart—is divided into twelve months, while the Mayans segmented time differently; some calendars were divided into 20 (like the number of fingers and toes) and others into 13 (like the number of joints in the human body, according to the Mayans). The moment when the great cycle of 5200 years ends and the new cycle is about to begin, that precise moment when an exhausted temporality gives way to a new one, is transcribed as follows: 13.0.0.0.0.0.

These superimposed layers of interconnecting calendars is an immense work of conceptual engineering motivated by an ideological assertion that is now shared by digital civilisations: reality is made of algorithms, reality is data that can be revealed and programmed. But, far from Silicon Valley and Shenzhen, far from the ultra-specialisation of contemporary science—which allows for the deepening of knowledge but hinders interconnectivity between different specializations—the Maya algorithms manifest themselves as the very need to connect phenomena: the movement of the stars, the ecological cycles of the planet, the social fabric of communities, and the biological units known as human beings.

The loss of memory of these scientific epistemologies thrown into the colonial bonfires for three centuries had been so overwhelming that when the ancient Mayan cities were rediscovered in the 19th century, many ventured to claim that this must be a civilization imported from another continent. According to these pseudo-historical accounts, this was the work of "wise white navigators" who had come from the Mediterranean in ancient times. These immense architectural complexes could not be—many of these European and American "experts" claimed—the work of the impoverished "Indians" whom they hired as porters for their "illustrious" scientific expeditions. The "grand" Western account had no memory of these other histories of science; that account ignored that Mayan astronomers would have identified the transit of Venus, centuries before their Asian and European colleagues—the celestial event that a glyph at Chichén Itzá seems to encode. And, it would be another few centuries before, elsewhere in the world, someone suspected that the transit of Venus might be the key to accurately determining the distance between the Earth and the Sun.

The British astronomer Edmond Halley—famous for predating the passage of the comet that later bore his name—devised a way of calculating this distance. His method requires at least two observers at two distant locations—although both must be on the same meridian strip, such as Japan and Australia—who observe the transit of Venus: the passage of that planet as it simultaneously aligns with the Earth and the Sun. These two observers time the duration of the astronomical event. The data collected by the two observers are then put together to obtain the distance between the Earth and the Sun by means of triangulation and various mathematical calculations.

The method devised by Halley is pristinely described in Covarrubias's book; the Mexican astronomer makes use of a broad pedagogical, rhetorical approach, more comprehensible to a wider public than the strictly scientific. In reading his book, one senses Professor Díaz—with his prominent moustache overflowing the corners of his lips—discoursing in front of his students, walking in the classroom, with the sound of his footsteps like a frequent percussion accompanying his voice.

In Yokohama during the month of November, the Mexican expedition prepared to build two wooden cubicles with solid stone floors to serve as astronomical observation posts. Both stations are located within the city but several kilometres apart, one in the Nogue-no-Yama sector and the other in the hills that the British had named the *Bluff*. They have chosen to make observations from two points, instead of placing all their instruments at one point, because they can then put Halley's method into practice—though without absolute precision, since this method ideally requires the second observatory to be elsewhere in the world on the same meridian strip as Japan—so that they do not have to wait for one of the other international expeditions, which did have observing stations in some distant country, to share their data with them. France, Australia, Germany, India, Russia, the United States, England, the Netherlands, and Italy set up telescopes along the meridian; dozens of astronomers scattered from Siberia to New Zealand ready to record the transit of Venus in front of the Sun.

The day is coming, the transit will take place on 9 December 1874, and Covarrubias is worried about the weather. The days are very cloudy and clouds happen to be the most fearsome opponents an astronomer has. He was also concerned about the political climate in Mexico; the press and many politicians had opposed the government investing the money needed for the expedition; returning home without bringing back the promised results would mean serious problems for all concerned.

****

In the night sky, after the Moon, Venus is the brightest luminous body recognised by the human eye; but observing its transit is only possible during the day. It is a reverse visual exercise and a challenge for the ocular iris. In front of the immense roundness of the sun, Venus is transformed into a tiny shadow, a black dot that seems to burn, a beetle embraced by nuclear fusion. Four billion years ago Venus was probably a place with ideal conditions for life to germinate, but today it is a scary place, the atmospheric pressure on its surface is 90 times more intense than on our planet, the temperature exceeds 400°C, and a rain of sulphuric acid corrodes everything it touches.

The Soviet space programme (CCCR) had clearly dominated the space race until 1966. That year they succeeded in landing a probe—kamikaze-like—on the surface of Venus, becoming the first human-made object to reach the surface of another planet; but 1966 would also be the year that death would pay a sudden visit to Sergei Korolev. The body of this virtuoso aeronautical engineer was about to turn into a chemical soup of which only bony sedimentation would remain. Korolev had led all of the CCCR's successes, from the delivery of the first animal to the first human into space. The sudden death of the "chief designer"—as he was known since his name was a state secret—would mark the decline of the Communist Party's stellar ambitions. A few months before his death Korolev had contemplated that in 1980 it would be feasible to send a manned voyage to Mars, but in the 1960s the moon was the major scientific and political goal to pursue. It is little known that there was a third contender in the special race, someone with the same ambitions as the "chief designer". His name was Edward Makuka Nkoloso, a school teacher in Zambia who for ten years led his country's space programme with the aim of sending "afronauts" to the Moon and Mars; moreover, he had ambitions to do so before the two powers succeeded. His plan failed as he was not allowed to test the rocket he had built. Perhaps because of the entrepreneurial fibre of this programme and his defiance of the big science-bureaucratic apparatus, it is fair to consider his project the predecessor of SpaceX and Nkoloso the predecessor of Elon Musk.

NASA's manned moon landing in 1969 ultimately confirmed the new American leadership. The CCCR then set about searching for other extraterrestrial destinations to generate the notion that the "conquest of space"—a slogan with a distinctly colonial spectre—was still in contention. Reaching Venus properly, capturing images and data, was the immediate goal.

Finally, in 1972, the Soviet space programme became the first to successfully land on Venus. The probe maintained contact with its base on Earth, sending back information on the atmospheric conditions and chemical composition of the soil of Venus. After 8 hours, the spacecraft was wiped out by the acidic crusher that is the Venusian ecosystem.

Forty years later, it was to be the Japan Aerospace Exploration Agency (JAXA) that would send an artificial satellite to orbit Venus. The Akatsuki probe is the only human-made object orbiting that planet, although one day its solar cells will eventually be ruined by space radiation. It will then begin to descend and once it passes through the acidic clouds it will plummet and begin its disintegration process. Perhaps in the near future JAXA will send space mining missions to Venus; it is possible—today the Japanese are leading the way in asteroid mining; they use small robots that move by jumping rather than on wheels. These robots photograph each other as if this were their summer vacation while drilling deep holes in the metal mass of the asteroid to extract what they are looking for. These images of their strange holidays are sent back to Earth. Sometimes the actions they record visually are understandable, but sometimes they produce an eerie silence in the large room where Japanese engineers and astrophysicists gather to observe them.

*****

Francisco Díaz Covarrubias woke up very early on December 9th, 1874. It is a cold morning and the clouds that in previous days were scattered in the atmosphere are now absent. The luminous halo of the sun is beginning to appear; its light travels to 300,000 km/s to meet the astronomer's corneas, He feels as if he is the first human to have ever seen the sun, to have discovered it. He writes, "The great centre of our world appeared on an entirely clear horizon, and never received a greeting so full of gratitude as on that day".

******

Agustín Barroso, the expedition's photographer, is already getting his equipment ready. It is almost a miracle that he managed to get everything he needed, as they had barely had time to prepare before their departure from Mexico. Very tall and thin, his movements are "a mixture of clumsiness and elegance" as Bulnes sometimes joked, and he is also dedicated and sparing, "Barroso is the most pessimistic or realistic among us" according to Covarrubias. Perhaps his colleagues perceive him as such because of the antipathy this photographer/photo-engineer feels for the political class, whereas both Bulnes and Covarrubias were to devote much of the following decades to pursuing political careers, and to being persecuted themselves.

Barroso had had to wait until he reached San Francisco to be able to buy the camera that he would use in Japan; "the violence with which the expedition had to set out because of the short time available to reach the final point of its destination," wrote Barroso in his official report, "meant that in Mexico it was barely possible to gather and pack the indispensable astronomical instruments; but it was not possible to have special photographic apparatus built, or even to prepare for it with the most common ones".

The invention of photography in the 19th century was to introduce substantial changes in astronomy. One of these is that once the image has been developed and fixed, the photographic plates could be minutely studied by providing measurements of the luminosity of the stars and the distances between them. These immutable astronomical images were a revelation; a revelation that was to cause a transcendental rebellion. The meticulous work of analysing the photographic plates was mostly carried out by women: their scientific contribution was to overturn old and heavy paradigms.

Running thousands of detailed measurements on countless photographic plates, computing and comparing them with others, requires remarkable data processing power. In 1875, the Harvard Observatory for Astronomy (HCO) hired three women—Anna Winlock, R. Rogers, and R. Saunders—to devote themselves fulltime to this laborious mission. Many others were to join in the following years, as the women's work "optimized" the HCO's budget, as they were paid only a fraction of the salary of their male colleagues, who were more concerned with operating the telescopes and obtaining the photographs of the night sky that were later studied. By 1920, forty "computers"—as they were known—were working at the observatory.

There is a story of which much remains to be told; the role that labour exploitation and discrimination have played in the development of astronomy. In this respect there is a straight line to be drawn between the Harvard astronomical observatory and the one that, centuries ago, was run by Tycho Brahe. In the first half of the 17th century, Brahe was the most famous astronomer in the West, and the Danish had given him an island to be the ruler of its few hundred inhabitants, and to set up the observatory he had been planning for years. The inhabitants of the island of Hven had to pay tribute to him by working on the construction of the observatory without pay and, under pressure from the astronomer, they were forced to pay him to build the observatory. They were also forced to double their harvests in order to subsidise the large expenses of the scientific centre. Sophia Brahe, Tycho's sister, was his helper but also independently conducted observations that added considerable amounts of data to the immense astronomical records being compiled on the island.

One hypothetical day, Sophia would have walked through the garden she herself designed outside the observatory; the garden begins in Hven and, like a constant reflection, perhaps like a folding of space-time, its green paths extend all the way to Harvard. Its concentric orchards are walked by the "computer" Henrietta Leavitt who discovered that stellar pulsations could be used to measure the distances between distant galaxies. The footsteps of Cecilia Payne also run through the orchard; she is the first person to obtain a PhD in astronomy at Harvard and discovered that the sun is composed mainly of hydrogen. If this timeless garden were to extend far to the east its paths would lead to China where Wang Zhenyi would be found conducting experiments to understand eclipses—never imagining that hundreds of years in the future a crater on Venus would be named after him. And, to the west, some path would flank biznagas, toloaches and peyotes to the site of Maria Francisca Gonzaga, the Criollo astronomer who in 1757 published in Mexico an almanac with accurate predictions of eclipses. For a long time the contributions of these and many other scientists were to be relegated to the fringes of the official canon. Exclusion, as a form of symbolic violence, can also appear in physical, embodied form, even to the point of murder, as in the case of the most famous mathematician of the "ancient world", Hypatia, who was dragged, stoned and quartered by a Christian mob.

The construction of scientific knowledge takes place from the claim of objectivity and neutrality, the idiosyncrasies of which Virginia Woolf describes as follows: *"It seems that science is not asexual; / 'science' is a man, it is a father, and it is also tainted. / Tainted science finds measures to suit itself".* These "contaminated" prejudices appear in several pages of Covarrubias' memoirs of the 1874 Mexican expedition. The author declares his "worries" about the new attitudes of fervent independence that he sees around him. On the way to Japan, the expedition had to cross the United States from coast to coast; the train journey took several days. At night it was possible to spread out small beds in which the passengers could sleep. The astronomer notes that "when I saw so many young and beautiful ladies who, when the hour of rest came, went quietly to their beds to undress under the shelter of a curtain, and this in the middle of the desert, surrounded by perhaps unknown men, I could not help remembering the times, not yet so remote, when a modest lady would have shuddered with horror at the thought of sleeping confidently next to a strange man", and he goes on to state that "in few countries are women so highly regarded as in the United States, for this regard reaches such a degree that it sometimes degenerates into exaggeration, turning some individuals of this beautiful half of mankind into real tyrants, demanding and malevolent, or into beings who aspire to occupations, positions and rights that are in every way incompatible with the obligations imposed on them by their sex, and who will always be rejected by reason and philosophy". None of the international missions observing the transit of Venus in 1874 included women among their participants. But if these expeditions extended that murky legacy of exclusion, on the other hand they succeeded in catalysing the deep human curiosity that embodies the luminous, inspiring side of science.

On December 9th at the observatory installed on the hillside of the Bluff is a Japanese intern—a young engineer the Mexicans call Mr. Koe—who has arrived with that morning's newspaper. He shows Barroso the front page of the newspaper, where there is a half-length portrait of Pierre Janssen, director of the French expedition in Japan, who appears in the picture holding in his hand a technological first that today is considered a precursor of cinema: "the photographic revolver". Barroso asks the intern "what's this machine?", who, after running his index finger over several paragraphs of the printed article, finally replies in a Japanese-British accent, "it takes many pictures very fast". At that moment Barroso hears a whisper in his left ear—it is Francisco Jiménez, one of Mexico's most prominent geographers who is also part of the expedition; "there are only a few minutes left before first contact begins—which is what they call the moment when Venus is observed entering the solar circle—

you'd better tell the kiddo to go and sit down". Jiménez immediately snatches the diary from him, his eyes stopping on the printed portrait. He looks at it fixedly but without giving away what he was thinking at that moment; "motherfucking Janssen".

Since their arrival in Yokohama, the Mexicans had sent several messages to the Frenchman to establish an official line of communication, but he did not and would not answer them, only allowing one of his delegates to maintain a minimal telegraphic exchange with the Latin American expedition. Barely a month after the transit, in an absurd occurrence, Janssen himself sent them a telegram requesting their collaboration in carrying out some additional measurements from the two Mexican observation posts; they sent him their apologies because they had already packed the instruments. Jiménez would have done well to save some of his annoyance for later, because they would have had more of it in Paris. When the Mexican expedition headed west to return home, crossing from Asia to Europe, they stopped for several days in Paris where they were to be ignored by their colleagues at the renowned Parisian Astronomical Observatory, who—disregarding protocol—did not receive them during their visit.

If Janssen had reciprocated, the Mexican expedition might have learned that there was another Latin American astronomer in Japan. The operator of the "photographic revolver" was a Brazilian, Francisco de Almeida, who had been sent by the Royal Observatory of Rio de Janeiro to study in Paris and would later be part of Janssen's expedition.

The Rio observatory was older than the one in Mexico City that Covarrubias had directed, but the first to appear in Latin America—following European models—another, tall white structure that in 1803 rose higher than the bell towers of the Bogotan cathedrals. Today it appears dwarfed by the skyscrapers that abound in the capital but two centuries ago it was the closest place to the sun that Colombians had built. Its first director, Francisco José de Caldas, was to join the insurrection against the Spanish occupation but was to be imprisoned. Legend has it that just before his execution he asked for clemency. One of his executioners reportedly replied "Spain does not need wise men".

****

The expedition obtains impeccable observations of the transit of Venus, and that same day Covarrubias sends a telegram to Mexico, in English, trying to prevent the telegrapher from making mistakes:

To. D. Sebastian Lerdo de Tejada, President of the Mexican Republic. Complete success in the observations. Please receive my most sincere congratulations.

Minutes later it is the astronomers who receive a written message:

The governor of Kanagawa presents his compliments to D. Francisco Diaz Covarrubias, President of the Mexican Scientific Mission, and requests the pleasure of his and his associates at dinner on Friday.

In the following days they were visited by British and Russian astronomers and invited to many events with Japanese ministers and scientists. But they received no reply from Mexico, which worried them a lot because the struggles for political power made the social life of their country unpredictable. They decided to continue with an idea that was already part of the plan they had discussed during the weeks of their trip enroute to Japan: to be the first in the world to report on the observations and conclusions of the Venus transit of 1874. This haste was due to their suspicion of a possible political trap, fearing that members of parliament or journalists opposed to the government might accuse them of having copied the results of the other international expeditions and thus cast doubt on their achievements; many were anxious to portray this scientific enterprise as a useless expense.

Despite the tensions in their country and the "ignoring" of their French colleagues, the Mexican expedition achieved its goal. Just two months after the transit, Covarrubias presented a paper at the Congress of Geographical Sciences in Paris, explaining in detail the records and mathematical calculations obtained from the observations in Japan. The presentation took the international scientific community by surprise; in such a short time these "unknowns" had managed to compute and report what would take months, and in some cases even years, for the other expeditions.

It is established that the long-sought distance between the Earth and the Sun is 149.59 million kilometres.

Two days after the Paris congress, they finally received a reply from their country. Probably due to some mistake by the telegrapher in

Yokohama, that first emotional message had not reached Mexico in time. The president replied:

The whole country thanks and praises you for your outstanding scientific work.

And it really is like that, when they return to their country everything is a party. They are celebrated in the streets, in the chamber of representatives, and also by the press. The enthusiasm is so widespread that one of the traditional pulquerias in the capital changes its name to "El Tránsito de Venus".

But since there is only one step between celebration and oblivion, the country was soon to return to dealing with its most urgent problems, of which there were many, and that event also represented a return to "normality": the constant dredges of Latin American politics that monopolized public discourse. Although in the decades that followed, Mexican academics made several efforts to ensure that this history did not disappear from social memory, in reality the country almost completely forgot this astronomical expedition. It would only be a century later that this memory would achieve a permanent social synapse, although this was not to happen in Mexico City but in Yokohama where in 1974 a monument was built to commemorate their 100th anniversary. The shape of the monument is non-human, it is a large stone, beautiful as all stones; one of its faces is flat and contains a text carved in low relief: it tells of the astronomers' visit and their mission. It will take about 10,000 years for the rain and wind to erode that text until it is unreadable.

A few meters behind the monument, the children of a nearby school have set up a mysterious metal object. In reality it is mysterious only to the attentive passer-by, for in the shadow of that large stone it can easily go unnoticed; its 78 cm height is quite inconspicuous. The same curious passer-by will find that the metallic object is a "time capsule" containing messages written by its authors to be read a century from now.

****

The next transit of Venus will occur on December 10th, 2117
Starting from 06:22 (Mexico City local time)

TROUBLE AT THE SYSTEM-ENSEMBLE BOUNDARY

Nolan Oswald Dennis

Prescience

A central issue. Science relies on a precise distinction between scientific and non-scientific knowledge to stabilise its claims to a uniquely superior form of knowledge based on rationality, empiricism, objectivity etc. However, philosophy of science has consistently struggled with determining precisely what the terms of this distinction are. Since at least Thomas Kuhn's intervention in the 1960's,[1] it's been increasingly apparent that, whatever the distinction between scientific and non-scientific knowledge might be, we primarily understand scientific breakthroughs (the moments which inaugurate an item of scientific knowledge) in retrospect. The line between science and non-science is at best an historical artefact. A difference which is knowable only after the fact and in relation to whatever prior knowledge some scientific breakthrough replaced. Treating the distinction between science and non-science as teleological rather than ideological underlies a political manoeuvre in western science which tends to obscure (and dismantle) the existence of "*many different kinds of science*".[2] Not simply different kinds of scientific fields (although the notion of a unitary science is undermined by deviant practice across fields, disciplines, even labs).[3]

1 See Thomas Kuhn, The Structure of Scientific Revolutions, 1962
2 Feyerabend, Against Method: Outline of an Anarchistic Theory of Knowledge, pg 2
3 Galison & Newman, Interview with Peter Galison: On Method, pg 6

But different sciences which arise from different peoples, different contexts,[4] and what Sylvia Wynter calls rhetoric-symbolic formations of the human subject - that is, different ordering epistemes. Any definition of scientific practice is inseparable from "the process of human evolutionary epistemology/modes of self-troping".[5] Scientific practice is then a particular boundary-defining coordinate in a normative ensemble of power.

Science puts into play the classificatory principles that bond a system-ensemble, basically a collective of legitimate Subjects (the trans-subject entity), through instituting collective systems of behaviour and ordering modes of knowing. The distinction between science and non-science (what Ian Hacking blithely calls bad nonsense)[6] effects an autopoietic whole in which other forms of knowing (and knowing Subjects) gain legitimacy only by *not* making any claim to scienticity. By accepting their non-scientific nature they are permitted legitimacy. Claiming things like intuitive knowledge, cultural knowledge, mythological, ethnographic, etc. allows these modes of knowing to persist by verifying the systemic code of the system-ensemble. As long as these ways of knowing do not call themselves science they are legitimate pursuits: recreation, history, hobbies, therapies, spiritual practices. So-called nonscience demonstrates its integration as part of the system-ensemble by respecting (and recreating) the system-ensemble boundary. Simply, to have a space, know your place. However, the history and future of scientific knowledge run contrary to this imperative, emerging precisely at the violation of normative codes: embracing heresy, abandoning common-sense, transgressing boundaries, acting irrationally, prayer, meditation, revelation.

It is relatively difficult to determine at any given moment whether the work of a scientist is producing sound scientific knowledge. According to Kuhn's schema,[7] *normal science*, the daily work of scientists within a stable theoretical framework - what Wynter calls a grammar of regularities[8] - mostly consists of piling up anomalies. These anomalies accumulate as more and more ways a theoretical framework fails "to mesh with facts about the world"[9] emerge. This is

4 Feyerabend, Op.cit, pg 3
5 Wynter, The Ceremony Must Be Found: After Humanism , pg 23
6 Hacking, Representing and Intervening, pg 3
7 normal science, crisis, revolution, new normal science.
8 Wynter, Op.cit, pg 38
9 Hacking, Op.cit, pg 7

well and proper. In a normative account, growing irrelevance compels a rewriting of knowledge.[10] Science, as a matter of course, undermines its own conventional assumptions and engenders crises. When enough anomalies accrue, normal science produces a crisis. Science becomes, on its own terms, no longer sound. Kuhn describes what happens next as a scientific revolution. Paradigm shift. A radical intervention into the grammar of regularities changes scientific consensus and produces a new normal science. The old normal science is relegated to the realm of non-science (think heliocentrism, phlogiston, energeticism, etc). A new set of stabilising assumptions is established and scientists begin the task of piling up anomalies again. Neat but misleading. Wynter reminds us that scientists are themselves part of the science they practise. The anomalies accrued in the process of *normal science* are not easily discarded by scientists because *normal science* is the representative practice of an epistemic order. The assumptions that stabilise a scientific paradigm are intertwined with the assumptions that stabilise a social paradigm. Rationality, an epistemic order, is a social relation which is "normative precisely because the structure of [scientists'] lived experience is isomorphic with the representation that the order gives it".[11] Since the anomalies are features of the theory, the user of theories cannot normally perceive them as anomalies. The user of the theory benefits from the existence of the theory. The user of the theory emerges alongside the theory. To discard the theory is therefore to discard the theoretician alongside. Or at least transform them. The anomalies are built in. As Kuhn says: every theory is born refuted. The notion of science as a self-transformative practice, moving from revolution to revolution, is more of a metaphysical hypothesis. Instead of accounting for radical practices, this view of science allows us to preempt the metamorphosis of rebellion into conservatism. Instead of a straightforward break with old forms, it suggests a higher unity (existing relations of power) in which radical knowledge is always already in service of, and constitutive of normative practices. Revolution demands something more.

In the normative course of practice, old science becomes non-science and a new science enigmatically arises (from where exactly is still, anarchically, unclear). Framed in this way, the distinction

10 Wynter, Op.cit, pg 22
11 Ibid, pg 39

between science and non-science seems relatively mundane. In this scheme, it is natural for scientific practice to render the object of its study obsolete. This is called Kuhn-loss, the inability for scientific knowledge to preserve insights from preceding paradigms.[12] In small and large ways science is seen as a self-destructive endeavour which is paradoxically self-preserving. Scientific practice is the destroyer of science which leaves in its wake more science. In spite of this seemingly evolutionary model, Kuhn's critical insight is that scientific knowledge is not cumulative but revolutionary. It consists of breaks across which no continuity is possible. Kuhn calls this a paradigm shift, a new disciplinary matrix which is incommensurable with the world that came before it. "A new theory is a new language"[13] which converts or displaces the speakers of the old language. Today's non-science is tomorrow's science, today's science is tomorrow's nonsense.

Scientific practice emphasises particular forms of rigour. Measurement, observation, annotation, experimentation, accuracy, precision, regulation, classification etc. Seen from afar, this continuous construction of procedures, methods and contrivances for the investigation and production of knowledge coheres into a common practice. A long challenged (if not overcome) relic in the popular history of science is the concept of the Scientific Method. A set of rules that supposedly govern the generation of new scientific knowledge. Ian Hacking calls this notion "the history of an error".[14] A generous description of a boundary-maintaining system in service of dysfunctional sovereignty. Partially derived from a Platonic over-regard for deductive knowledge in relation to the instability of historical, material evidence. Methodology misapplies various artefacts from scientific practice (regulation, procedure, classification etc.) as rules of a symbolic system which stabilises the distinction between science and non-science. Distanced from historical practice, this symbolic system relies on procedural differences (did you follow the correct steps) rather than differences of content (did you get an accurate result) to distinguish "bad nonsense" from "good science".[15] Indeed, following these rules is presupposed to

12 https://plato.stanford.edu/entries/thomas-kuhn/#PercObseIncoWorlChan
13 Hacking, Representing and Intervening, pg 12
14 Ibid, pg 2
15 Ibid, pg 3

guarantee progress.[16] Scientific method suggests a ceremonial order of truth. The elements of the method become seeking/avoidance protocols which interdict "any [other] ceremony which might yoke the antithetical signifiers and breach the dynamics of order/Chaos, through which the order brings itself into living being".[17] Paul Feyerabend calls this scientific chauvinism the idea that "What is compatible with science should live, what is not compatible with science, should die".[18] Since science, western science, and bourgeois science, are treated as indistinguishable; methodology, as a boundary-maintaining system, is simply a political extension of colonial logics of governance. Feyerabend highlights that the hegemony of western science is not due to *"insight in its 'inherent rationality' but power play (the colonizing nations imposed their ways of living)".*[19]

According to the Science Council the scientific methodology consists of seven elements: Objective observation; Evidence; Experimentation; Induction; Repetition; Critical Analysis; and Testing.[20] This method is in fact a model, it suggests an ideal set of components which science practice is *like.* Similar to. Aiming for. It is an image of practice which stands in place of the actual, unwieldy process of creating knowledge. It indexes a hierarchical principle in which a specific genre of rationality is the ideal form of cognition, superseding and opposed to, rather than complimentary with extra-rationality (intuition, inspiration, revelation). Wherever there is an indexation of parts, there is also the denial of others. In this scheme science resides in a performance of scienticity, a certain manner of work, an epistemological prescription[21] which resides in matters of good conduct. Natives must get haircuts and Western suits.[22] A collective system of behaviours ordered by the episteme - good manners, a presentation of respectability, disciplined action, altogether a performance of deference. To what? Wynter reminds us that a rational worldview is a reciprocally auto-instituted system-ensemble which brings a specific normative identity into being, that a disciplining episteme is also a classifying principle.[23] Adherence

16 https://plato.stanford.edu/entries/thomas-kuhn/#PercObseIncoWorlChan
17 Wynter, Op.cit, pg 27
18 Feyerabend, Op.cit, pg 36
19 Ibid, pg 3
20 https://sciencecouncil.org/about-science/our-definition-of-science/
21 Feyerabend, Op.cit, pg 12
22 Paperson, A Third University of Possible, pg 59
23 Wynter, Op.cit, pg 22

to the scientific method inaugurates the scientist as an ideological figure which permits one entry into the community of the rational.

Let me cut to the point. Methodology is a disciplining formula which demands normativity as a condition for rationality in "the colonial hope that the whiteness of the normative human can be extended to the very people who were premised as non-human, gender-deviant savages".[24] Assimilation. Hacking puts it simpler: "The pursuit of truth and reason will doubtless be organized according to the same social formulae as other pursuits such as happiness or genocide".[25] Scientific methodology presents a collective system of behaviour united by a set of abstract principles, the elements of scientific practice, which are redirected from their disruptive function as a means to knowledge toward a stabilising purpose as technologies of alienation. Knowledge generated in this manner is thus considered scientific because it has been produced through the application of the scientific method. Other ways of producing knowledge are rendered unscientific, regardless of their truth, rigour or correspondence to reality. That is, even if they produce scientific knowledge, non-methodological practices are unscientific means to knowledge.

The paradoxical outcome of methodological regulation of scientific practice is the mystification of scientific processes. Under the guise of rationality, a bureaucratically sanctioned identity arises—*the scientist*—whose behaviours are orientated by an ordering system called the Scientific Method. Structurally generated from the opposition between scientific and unscientific categories, *the scientist* must busy themselves with the decision-making processes of what to in/exclude. A mystic practice. With the ever-present fear that, should the boundary-maintaining dynamic of the Scientific Method fail, acknowledging the irrational elements already present in scientific practice might allow the inclusion of non-western knowledge practices on their own terms (rather than as scientific reconstructions).

Science constructed this way is only permeable to non-white, non-western cultures insofar as they remain external to the normatively White Euro-American intra-cultural space designated as science. Science proper. African science, Indigenous science, Asian

24 Paperson, Op.cit, pg 59
25 Hacking, Op.cit, pg 11

science, Black science, etc., are only allowed in as extra-scientific artefacts of various ethno-sciences in relation to Euro-American pre-sciences. Yesterday's nonsense in relation to today's science. Anything else would be fatally at odds with the historical entanglements of science as an extension of a colonial world project. The disavowal of external standards of knowledge is also a disavowal of other worlds. An attempt to fix the boundaries of a system-ensemble on procedural grounds, to freeze the world at the high point of Western hegemonic power. The over-coupling of presuppositions and knowledge suspends conditions for the growth of knowledge. Change. Faced with challenges to the founding structural oppositions of a western scientific order, desire for scientific method (more precisely a unified scientific method) suggests it is better to operate a model of research which masks prejudice and mystifies scientific practice than forgo the privileges granted by such an order. Until the revolution *really* comes.

Ana-logics in a non-aligned cosmos

1. *In thick black, I write* كذب المنجمون ولو صدقوا *(a popular Arabic saying misconceived by many as a verse from the Quran that translates as: "The stargazers [conjurers/fortune tellers] lie even if they tell the truth")...*[26]

2. In the mythology of physics there is a tree, an apple falling, a walking Isaac. This is a non-repeatable portion of science. Here the apple is analogous to the moon. The apple is like the moon, its fall from the tree is, unexpectedly, similar to a lunar orbit, at a squint it even resembles the moon. Convenient. According to Voltaire, this is "the first thought of his system of gravitation".[27] Of course, his first thought is also our first thought, accordingly. Newton called it his inspiration, the first breath of his systema. An apple, now a moon, a sun and its cousin stars. A galaxy amongst galaxies, after *galakt* meaning milk. A kitchen of short-lived theories. A few hundred years later the significance of this fruit tree is revised by a pair of contemporary scientists. Newton's first moment is retconned as his last: instead of an inspired 20-something year old we are offered an "elderly Newton [who]

26 Raafat Mazjoub, Scripting Stars, https://perpetualpostponement.org/scripting-stars/
27 Voltaire (1727). An Essay Upon The Civil Wars of France, pg 104

had forgotten that he had once posited it as an explanatory analogy".[28] The apple now comes after the knowledge. Rather than inspiring the physics the apple story is now *meta* Newtonian physics. In this version the apple tree is non-scientific (if not exactly non-sense). A lie to children conveniently shaped to fit the methodological model. This is a stratagem at an epistemic border. The negation of analogical reason is necessary for the construction of science as a methodologically consistent, rule-based practice. Analogic reasoning is necessarily unreliable and so revised as an explanation, not a method of discovery. Even when historically evident, it is safer to rewrite history than to tolerate an analogic presence.

3. There is no reliable formula to reproduce the incidental pairing that grounds analogical reason. A co-incidence. There must be a preconception of the unknown thing (an in-sight) in order for an appropriate analog to present itself.[29] Always the appropriate practice precedes the correct method. Determines it even. A sense of what might be appropriate to recognise a target and an analogue. Which is to say: meticulously construct a presumption, and then apply the presumption to a set of practices which, in their most promising form, are concerned with transcending presumption. Let's assume we always start halfway. Knowing something and not knowing something else. In an analogical structure, to know something means to put it in relation with something else. Poetic relations are important in analogical reason (see Newton's study of the Hermetic tradition), scientific analogy emerges from the complementarity of substance and signification, a material relation between observation and representation (and uncertainty).[30] Analogical reason is a relation of relations. Reckless too.

4. In /Xam mythology, the sharp light of an angry sun cuts the moon into smaller and smaller pieces until, mercifully, the sun releases the moon, a man, to grow again into his full magnitude.[31] Analogy is more-or-less reducible to a mapping function.[32] The condition

28 Gribbin & Gribbin, Out of the Shadow of a Giant: Hooke, Halley, and the Birth of British Science, pg 165
29 Lee, Analogical Reasoning: A New Look at an Old Problem, pg 3
30 Barad, Meeting the Universe Halfway, pg 3
31 Slotegraaf, Traditional Star Lore of Africa, pg 2

of any analogy is a pair of minimally related objects. An analogue and a target. The analogue is conceptually mapped onto the target such that the former reveals something about the latter. A revelation. A general condition of analogic reason is an explanatory or predictive function. The persistence of any analogic is a function of its usefulness. While mystery is built into analogical reason, as a revelatory method, analogue-building is a distinctly demystifying practice - it reveals. Via some unrepeatable process, a thing (the moon) is suddenly seen as if it is another thing (a man in trouble with the sun) in such a way that we understand the relationship between the two celestial bodies usefully. This knowledge forms according to its own spiritual historical logics irreducible to others without missing the point. A certain induction into the type of knowledge we seek is necessary in analogical reasoning. A counter-inductive attitude.[33] A visionary blur between two things.

5. As an ampliative mode of reason, the conclusions formed through analogical thinking are only supported to varying degrees of strength. Inconsistency is an inescapable feature of this manner of thinking. The aberrant practices of experimentation, the avant-gardist, queer, Afri-Indigenous, and altogether extra-ordinary histories of experimental life, are generally avoided in the scientific conception of experimental work. Instead, experimentation is reduced to its procedural and demonstrative dimensions. The unruly aspects of science are eroded in the interest of the formulaic which lends scientific practice an aura of tedium built on a diminished notion of rigour. An absolutely necessary but nonetheless insufficient condition for the production of any knowledge.

6. A system is any arbitrarily separated part of the world subject to a method of analysis. A system is distinguished from its surrounding environment by a boundary. Interactions at the system boundary determine the class of the system. In thermodynamics, open systems allow matter and energy transfer, closed systems allow only energy, and isolated systems allow no possibility of exchange with the surroundings. These are mostly imaginary conditions. Useful too.

32 https://plato.stanford.edu/entries/reasoning-analogy/#AnaCon
33 Feyerabend, Op.cit, pg 20

7. A straightforward account: analogies are heuristics for generating new knowledge, formulating solutions to problems, and generating more problems. Analogical reason is mysterious, it resists formalisation. Analogical reason is flawed, analogies increase credence (belief) without an increase in evidence. Analogy is contingent on the perspective of an unwieldy whole. Another system-ensemble awaits across the boundary. A meta-boundary ensemble. After the boundary another boundary, and so on. Borders work both ways, reciprocally produced by the fact of their surroundings. An environmental analogue. Surrounded by every unaccounted (unaccountable) thing. The unownable. The wild. The surrounding anti-cosmos resisting cosmology (the ordering system). The black before and before. Or better "the settler is portrayed as surrounded by "natives," inverting [...] the role of aggressor so that colonialism is made to look like self-defence".[34] A cage (after Faraday) is also a shield (after Faraday).

8. Fieldwork: A key characteristic of analogical reason is the use of previous knowledge, the already-known, as a guide to new knowledge, the totally-unknown. Model building. Consider the field. Micheal Faraday. The electromagnetic field. The force between. A land analogue. The farm. The plantation. The wave follows the field. An ocean analogue. The wake. Typhon coming on. A hand waving goodbye. Or "Hello, world!". Radiating energy. The spectrum precedes the field. The spect. The suspect. A ghost analogue. In the shell. Coiled. Ellison's invisible man.[35] There is a spectre haunting.

34 Harney & Moten, Politics Surrounded in South Atlantic Quarterly, pg 985
35 Ellison, Invisible Man, 1952

EVERYDAY IS A NEGOTIATION
Merve Ertufan

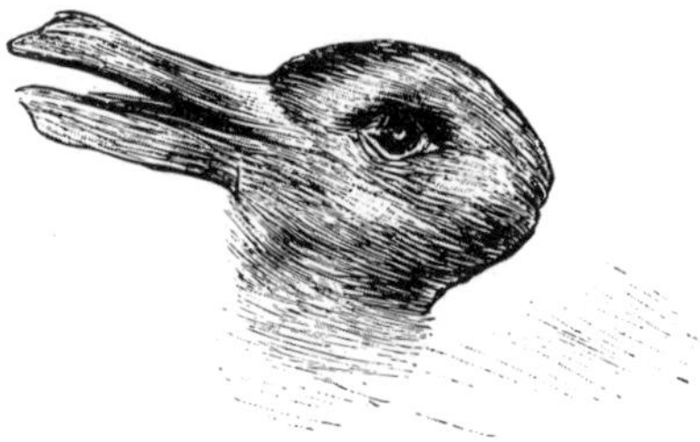

Jastrow's illusion, made famous by Ludwig Wittgenstein, asks whether we see a duck, or a rabbit, or a duck-rabbit. To this question, many might give different answers.

I would like to propose that if we ever had the impulse to grab this animal as it is in the image, we would all reach for the same area.

Alien Hand Syndrome:

The Alien Hand Syndrome is a rare condition in which one limb behaves outside of the range of control of the brain / the person. It acts seemingly on its own, without regard to the owner's wishes. The alien hand chokes, unbuttons your hospital gown, crushes paper cups, makes moves in checkers, fights your other hand, scribbles, or gropes for nearby objects. In its first appearance in medical journals in 1908, the neurologist and psychiatrist Kurt Goldstein describes the experiences of a 57 year old stroke patient, as she struggles with her left hand, stating that "the hand is not normal; it does what it likes", evoking an image of a hand that has its own desires.

Unlike a knee-jerk reflex, however, this involuntary action deals with goal-states. It is not a straight-forward kick -a simple action that the alien hand performs; it is a large set of complex actions such as grabbing and interacting with objects. It is as if the hand is a sub-personal identity that has a conception of the world and a will of its own. Not only does it appear as if the hand can distinguish between the object and the background, it also seems to know what to do with that object. Buttons are button-able, bottles are graspable, chess pieces movable, the zipper zippable.

And what is most remarkable is that the whole operation happens before intentioned action or linguistic articulation. The damage in the brain causes the alien hand to short-wire an already existing data stored in the brain, and just like that, it reveals how we cognitively engage with our surroundings. First, we differentiate between foreground and background, recognizing an object as a separate entity. Then we play with them, recognize the ways in which we can interact with them, name them, write instructions to tell other people how to interact with them, and then we combine or disassemble them, make new versions of them, and rename them. The epistemological structuring is complex, utilizing testimony, memory, perception, introspection and reason.

The ways in which we are able to interact with objects are called 'affordances'. What can an object afford to be used as? In an encounter with a handle, how do we know to hold it and turn it? The object itself doesn't actually include this information, but it carries signs that people can recognize from other more familiar objects, or it is within the range of movement we already know our bodies to be capable of. All these together make up a certain pre-existing set of social or personal codes. What's more, sometimes objects teach us to move our bodies in a certain way, and this information also finds its way into the pool of codes. So much so that even in the absence of the object itself, we can gesture and conjure an image off it. A fan, earphones, a lighter, etc... And other times, these signs expire: children today may not recognize the thumb and pinky sign for the phone. It depends on our body-image as well–for this, we need not think in such a normative manner. A handle may not be handled, but opened with a chin or a mouth. And it is this stored knowledge that the alien hand is able to access and make use of, while bypassing intention, casting doubts on certain inner workings of our decisions, as well as consciousness itself.

W-a-t-e-r / Shoehorn:

Suffering a disease as a 19 months-old toddler, Helen Keller becomes blind and deaf right before she learns to speak. Completely uninformed about how to linguistically navigate the world, she later writes about these early life experiences and the moment she learnt that things have names. She writes, "everything had a name, and each name gave birth to a new thought." The precise moment comes when she is 6 years old, when her teacher Anne Sullivan, takes her to a water pump and pumps water on Helen's one hand, and repeatedly spells "w-a-t-e-r" on her other hand, first slowly then rapidly. Helen describes, "I knew then that "w-a-t-e-r" meant the wonderful cool something that was flowing over my hand." She would later describe that profound shift as the time "when I was restored to my human heritage."

Interestingly enough, long before she set off into her new life with language, in her nonverbal state, she had figured out the door, the handle, the lock and key mechanism. And it is when she locks her mother in the pantry, the parents decide she needs education. They find her a teacher, who receives the exact same treatment when she arrives. Helen locks the teacher in her room and hides the key, prompting her father to use the ladder to get the teacher out through the window.

Madeline J., contrarily, is 60 years old when her path crosses with Oliver Sacks. In his book titled *The Man Who Mistook His Wife for a Hat*, Sacks tells the story of Madeline J., who is congenitally blind and does not know how to use her hands, even though they are in fact functional. She can't identify anything placed on her hands, she doesn't even have the exploratory use of her hands that infants learn in the first few months. Her hands are completely inactive, due to everything being done for her. She had never needed to use her hands, as everything was given readily to her. She calls her hands "useless godforsaken lumps of dough- they don't even feel a part of me."

The change starts with a bagel. Upon Sacks's orders, the nurses delay hand feeding Madeline. They place the tray in front of her, put the food within her reach, and each time make it a little later that they return and actually feed her. Until one day, fed up with hunger, she herself reaches for a bagel and puts it in her mouth. Her first manual act in 60 years, marking her birth as a motor individual.

An incredible leap in her agential capacities. This gesture also sets off another processive mechanism, that of pattern-recognition. After the bagel, there is rapid progress; she touches everything and anything, trying to connect those to the names of things she knows. She is of exceptional intelligence and literacy, and already has an imagination filled with images conveyed by language with words. However, since she had not used her hands ever, she has no tactile memory, no internal images of objects. Recognition is achieved "curiously roundabout", with inferences and guesswork.

For both of these persons, what they once had no access to later becomes a major part of their lives. Helen becomes a writer, publishing *The Story of My Life* in 1903. Madeline becomes known as 'the Blind Sculptress of St. Benedict's'. "The commonest objects delighted her—delighted her and stimulated a desire to reproduce them. She asked for clay and started to make models: her first model, her first sculpture, was of a shoehorn."

Tools, Sounds, Symbols:

After writing his most famous work titled *Tractatus Logico-Philosophicus*, there is a period in which Ludwig Wittgenstein retires from philosophy, and works as a gardener. He contemplates on becoming a monk for a while, but then he gets asked by his sister to design her new house. Wittgenstein is not the architect of the house; instead, he focuses on the windows, doors, and radiators, demanding that every detail be exactly as he specified. He takes one year to design the door handles, another year to design the radiators. It is only after working with proportions and precision to such a minute detail that it drives the locksmiths and the engineers to tears, he returns to philosophy. Christopher Benfey claims that it is the 'handle' that prompts Wittgenstein to return to philosophy, who writes all about handles and their relation to words in his next work *Philosophical Investigations*:

"11. Think of the tools in a toolbox: there is a hammer, pliers, a saw, a screwdriver, a rule, a glue-pot, glue, nails and screws. The functions of words are as diverse as the functions of these objects. (And in both cases there are similarities.) Of course, what confuses us is the uniform appearance of words when we hear them in speech, or see them written or in print.

For their *use* is not that obvious. Especially when we are doing philosophy!

12. It is like looking into the cabin of a locomotive. There are handles there, all looking more or less alike. (This stands to reason, since they are all supposed to be handled.) But one is the handle of a crank, which can be moved continuously (it regulates the opening of a valve); another is the handle of a switch, which has only two operative positions: it is either off or on; a third is the handle of a brake-lever, the harder one pulls on it, the harder the braking; a fourth, the handle of a pump: it has an effect only so long as it is moved to and fro."

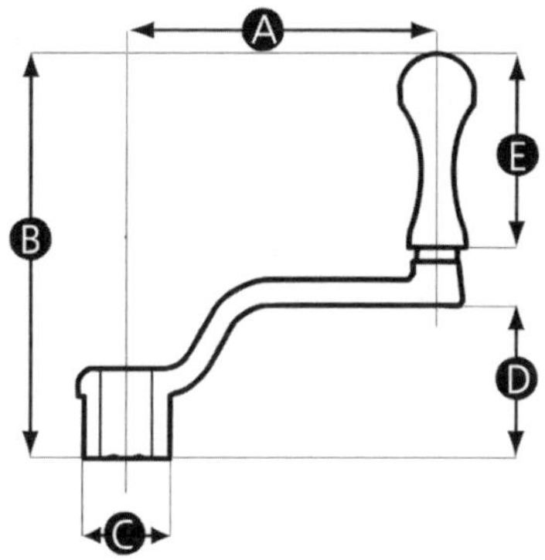

Ambidextrous crank: Offset Cast Iron Cranking Handle with Square Hole. Model 06 CR, manufactured by Rencol.

Yet, it's not just about the body encountering the handle. It's not that one navigates the other, where one is with intention and the other is a mere tool to achieve a goal. Instead, the tool and the body together generate a 'new'. So much so that at one point linguists were tempted to try and trace the beginnings of poetry back to early work songs:

"Every form of physical work conditions, already when it is performed by the individual but even more when it is performed by the community, occasions a specific coordination of movements, which leads in turn to a rhythmic organization and punctuation of work phases. This rhythm exhibits itself to consciousness in a twofold mode: first, it finds its expression in the pure sensation of movement, in the alternation of muscular tension and relaxation, and second, it finds its expression in objective form, in the perceptions of the sense of hearing, in the symmetry of the sounds and noises that accompany work. The consciousness of doing and its differentiation are connected to these sensible differences: milling and grinding, pushing and pulling, and pressing and trampling differ with respect to their specific purposes and to their own proper meter and tone. In the abundance and diversity of work songs, in the songs of spinners and weavers, threshers and oarsmen, millers and bakers, etc., we can still hear with a certain immediacy how a specific rhythmic sense, determined by the character of the task, can only subsist and enter into the work if it is at the same time objectified in sound."

Sounds of the production process come from both the tools and the users themselves. They come from each unique gesture, each unique tool and material; and sometimes, from the accompanying work songs. The causal bond between the repetitive physical movement performed by the laborer and the rhythmic sounds that come from working with tools result in an unavoidable phonetic repetition. Moreover, different kinds of work require and produce different movements and sounds. It is this differentiation, followed by categorization, and identification, that result in specialization. A process that eventually consolidates meanings and attaches them to signs.

Ernst Cassirer reflects on Karl Bücher's *Arbeit und Rythmus* (Work and Rhythm) in his four volume series on *The philosophy of Symbolic Forms,* and claims work as "the first time the rhythm felt by man in his own physical movements was, as it were, objectified." And even though he declares how simple phonetic repetition may be among "the earliest motives of language formation"; he unfortunately doesn't get swayed by the connection between tools and language. In his quest to demarcate the origin of language, he stands firm in his conviction of the individual and its centralized self as the prime originator from which words are produced, placing the "I" in the dead center.

The Turkish Shrug and the Anthropo-technical Injunction:

One can perform an ingrained Turkish shrug at any information that happens to arrive by the way of "-mış", instead of "-dı". Even on a grammatical level, these two past tense suffixes in the Turkish language make a distinction between information derived from hearsay and information from first-hand witnessing, pitting epistemic power of testimony against that of memory. The "-dı" at the end of a verb means "I witnessed that with my own eyes", whereas the "-mış" suffix translates to "I heard that from someone else", automatically imbuing the statement with doubt. It is an ontological doubt that saturates almost all information, because how much can one really witness? And so, how much can we really know?

Everyday is a negotiation for the Bayesian agent
(a.k.a the common man):
knowns / unknowns / known knowns / unknown knowns /
known unknowns / unknown unknowns

In Game Theory history, Frank Ramsey, Thomas Bayes, and Leonard Savage build up the figure of the rational agent. Ramsey dealt with the information we currently hold, and our belief in that information. I.e if we know that we should turn left from this road to reach our friend's place, how much do we trust that we remember correctly? Bayes was all about making the best decision with respect to the decision of all the others. I.e the two prisoners will avoid prison if neither admits guilt, but in order to make this decision they have to predict that the other will make the same decision. Savage put forward a model of rational agent based on the idea that we can predict the behavior of a decision of another rational agent by supposing that they will act in order to maximize the utility and avoid the worst possible scenario. Underlying this is the belief that people can be trusted to recognize and/or are capable of recognizing what is good for them.

Anna Longo bridges the inductive reasoning previously attributed to humans (organic intelligence), to contemporary AI (synthetic intelligence). Inductive reasoning occurs when one draws general conclusions by observing specific conditions and recognizing the patterns they supposedly exhibit. From specific to general. "The transformation of inductive reasoning which led to the automation of the production of predictive hypotheses via the statistical analysis of data... was brought on by the introduction of the use of probability, and in particular Bayesian statistics. This transition is what made a strategic conception of the machine learning process possible, in other words it is what allowed it to evolve. Predictions are selected based on the effectiveness of the decisions they back up, according to pragmatic criteria, of course. Mathematical game theory is therefore established in order to model the probabilistic procedure of inductive reasoning: dependable predictive hypothesis —the conclusions which everyone must consider true or rational— are those which permit a maximum amount of utility." When approached from the perspective of game/decision theory, in that case, the inner workings of how we deal with hearsay gets laid bare. The source of information is ultimately and utterly dubious. It travels by the way of words, it exists in words, in an abstracted state, molding and changing with every utterance, every tonation. Even with the use of cameras and recording devices, its ontological condition cannot be reversed. The rational agent that is called upon to make a decision with regards to the hearsay was not at the site of the event, and can never go back in time and be there. In that case, how can one judge the validity of the information and make predictions from it? Anna Longo warns that generalized conclusions gain more traction than their smaller origins, that they get made in regards to likelihood and utility, and that they not only disregard but also block many other possible outcomes. 'What could have been' is struck down with every decision. And in the case of hearsay, three questions emerge as we prepare to make a decision: "do I trust the channels that carried this information?", "is it in line with the previous information that I already hold to be true?", and "does it serve me to trust this information?" Since I am also a rational agent, I should act in accordance to what permits a maximum amount of utility.

And in accordance with whatever judgement is made, there is a further set of decisions to be taken: "do I build upon this information?", "do I dispute it?", or "do I leave it on the back burner?" There is a phrase in Turkish: "hang it as an earring." It dangles right outside your ears. Not yet in, but acts as a reminder or a filter, possibly to be proven useful in the future.

A Turkish shrug can appear at any moment of this process. But more importantly, it can appear at the precipice of a hearsay, cutting down its initial power immediately. In some ways, it is a denial of responsibility, a shrug that says "I don't know, I wasn't there." However, it is also an acceptance of the limits of our cognitive capacities, as well as a refusal to engage with the presuppositions, terms, and conditions that are imposed on us with the new information. A disengagement. A slight shoulder act, one that negates any other act from any other body part.

Rubber Hands, Macaque Monkeys:

So as to not confine 'hands' to their role in the capitalist order, nor to their part in the formation of language, it's good to remember that sensations accompany each touch. In their text titled *On Touching—The Inhuman That Therefore I Am*, Karen Barad writes, "When two hands touch, there is a sensuality of the flesh, an exchange of warmth, a feeling of pressure, of presence, a proximity of otherness that brings the other nearly as close as oneself. Perhaps closer. And if the two hands belong to one person, might this not enliven an uncanny sense of the otherness of the self, a literal holding oneself at a distance in the sensation of contact, the greeting of the stranger within?"

A curious experiment, however, from the University of Pittsburgh in 1998 succeeds in confusing ten subjects into *feeling* that a rubber hand is their own:

Experiment 1: "Rubber hands 'feel' touch that eyes see" (DIY version):

Ingredients:
2 people
1 table
1 screen to hide the actual arm
1 life-sized rubber hand (somewhat believable)
2 paint brushes
Optional: 1 blunt object

Instructions:
One person becomes the subject of the experiment. The other person, then, becomes the non-subject and performs the experiment. You can take turns.
The subject rests one of their hands on the table.
The screen is placed between the eyes and the hand, so as to remove the actual hand from the line of sight.
The rubber hand is placed on the table directly in front of the subject, in a believable position. (One can also facilitate this process by perpetuating a visual link from the shoulder to the hand to help the sense of ownership i.e using a sweater to invoke a visual resemblance of an arm.)

The non-subject starts to repeatedly stroke the actual hand and the rubber hand both, synchronously.
In about 60-90 seconds the subject should start to adopt the rubber hand as their own. This can be tested out by stroking only the rubber hand and asking the subject if they feel their hand is being stroked.
Optional: Smash the rubber hand with the blunt object, to see if the subject reacts as if their actual hand is being smashed.

Thomas Metzinger, a philosopher of mind and consciousness, recalls the moment right before his mind started to identify a rubber hand to be his own, which came along with a strange tingling sensation in his shoulder. He had exposed himself to an experiment from decades back, where two psychiatrists named Matthew Botvinick and Jonathan Cohen managed to trick their subjects into experiencing a rubber hand as their own. The fooled rational agent.

Metzinger would later build his 'phenomenal self-model' hypothesis on this exact experiment. A bodily self-model which aids us in experiences of ownership, of first person perspective, and of a long-term unity of beliefs and attitudes. This model is practically essential in interacting with the world with/through the body. It creates a sense of a personal border via an image of a singular, individuated, whole person within a world where we feel our limbs as ours. "This hand is mine, and I'm reaching for the apple on the table, because I'm hungry." It is a self-model that creates a strong sense of 'mineness', agency and intentionality. However, what the rubber hand experiment reveals to Metzinger is the utter malleability of the phenomenal self-model. In this nod to the cybernetic mind model, the eyes see what the hand can't feel, but since the brain thinks the hand we see is ours, it feels what it sees.

Atsushi Iriki, in his 2019 lecture at the International Convention of Psychological Science, proudly declares how he used the best technology thirty years ago for when he taught macaque monkeys how to use endoscopy: the smallest camera of the time, and the world's first commercial liquid crystal TV:

Experiment 2: How to get macaque monkeys to use endoscopy tools (non-DIY version):

***Setup:* body is pinned, can only move eyes and hands.**

Steps:

Step 1: A snack is on the table, visible to the naked eye, but out of reach. Use a rake to reach the snack. Reach is extended by the rake.
Step 1.5: use the short rake to reach for the long rake, in order to reach for the further away snack.

Step 2: A snack is hidden behind a hill-like curved structure at the end of the table, blocking line of sight. The snack is again out of reach. Use a stick with a mirror at its end to figure out the placement of the snack and flip it from behind the curve. The vision and the reach are extended with the mirror and the stick.

Step 3: There is a panel blocking the table. A snack is on the table and again not visible. Attach a camera on the rake, and a monitor to figure out the placement of the snack. The snack is only visible through the monitor, and the camera is moved by hand. Once the snack is visible, use the rake to reach the snack. Vision and reach are mediated through digital technology, much like the endoscopy tool.

I lean left with the Ps4 controller in my hand, playing Gran Tourismo, taking a sharp left turn.

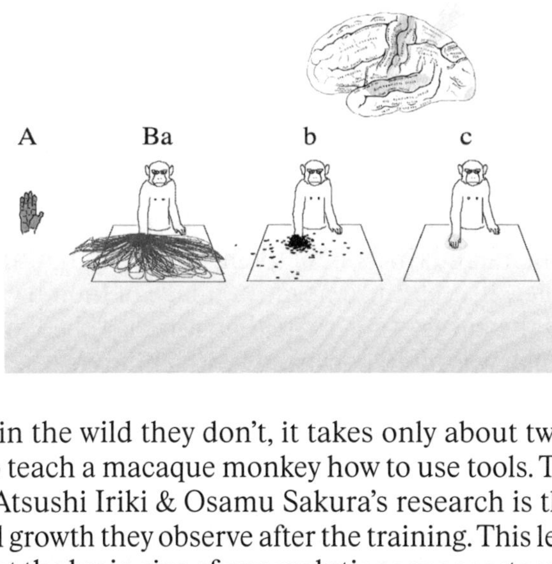

Although in the wild they don't, it takes only about two weeks of training to teach a macaque monkey how to use tools. The key discovery in Atsushi Iriki & Osamu Sakura's research is the correlatory neural growth they observe after the training. This leads Iriki to look back at the brain size of our evolutionary ancestors. He points at a breaking point in the increase of cranial capacity right around Homo Habilis (from 2.8m to 1.65m years ago), after which the brain capacity starts to grow exponentially—'Too Rapid Hominid Brain Expansion' as he calls it. Also known as 'handy man', Homo Habilis is the first of our ancestors known for tool use and language. Iriki attributes this change to be the historic turning point of human intellect, a turn that was first recognized by the anthropologist and archeologist André Leroi-Gourhan. Leroi-Gourhan associates these structural changes with bipedalism in his *Gesture and Speech* book. In short, as human ancestors started to walk on two legs "it released forelimbs from locomotor functions, thereby allowing them to be used in tool production...the face from feeding and manipulative functions, thereby allowing structural specializations for speech... overall causing changes in brain size and organization."

What Metzinger takes from Iriki & Sakura's experiments for his 'phenomenal self model', is the fact that the monkey cognitively extends his attention to include the rake. "You can engage in goal-directed and intelligent tool use only if your brain temporarily represents the tools as part of your self." In simple terms (or from the first person perspective of the bodily experience): when we write, we don't look at the tip of our fingers, we look at the end of our pen. "Any successful extension of behavioral space is mirrored in the neural substrate of the body image in the brain. The brain constructs an internalized image of the tool by assimilating it into the existing body image."

The body image, as it is an image, allows itself to be seen as well. Here is where we get the meta-self: "the capacity to distance yourself from your bodily self—namely, by beginning to see your own body as a tool."

When people started to look at their hands, the mouth was involved. They held their hands in front of them, filled their mouths with pigment and blew it, creating a negative image of their own hands on the walls of the caves. An activity that stimulated cognitive development, which was often performed with children.

Idle hands find something to do:

Darian Leader–reflecting through a vast amount of resources derived from early development of infants, modern day contraptions, and daily usage of hands–cautions against idle hands. "What happens when the hands are kept from their business? If a child is separated from their Lego or loom bands, or an adult from their mobile or console, very soon the body takes its place. Hands scratch, rub, pluck, press and pull. The overflowing agitation of the hands finds its way back to the body, a return that hand technology defers and inhibits. There is a circuit here that pulsates with a ferocious energy."

Albeit unnecessarily presented as a danger, there is a certain acceptance here: hands -even non-alien hands- continuously move. They don't have to, nor do they need to, but they just do.

So then, what would happen if we were to redirect the attention of the hand to its mirrored counterpart in the rubber hand experiment? Would touching 'your' rubber hand with your other hand, dissolve the illusion of ownership? Can the fooled rational agent, the one who had so easily taken up a rubber hand as its own, realign his phenomenal self model with the help of his other hand? And if indeed he can, how important is standing still for the rubber hand experiment?

No-brainers:

Design-wise, an object should communicate how one is to interact with it without the need for instructions, instinctively. A door handle should indicate with its size, shape and color—without the help of instructions—how it should be handled. It should interact with our sense of bodily agency, physical memory, and spatial appreciation. A well-designed-object should trigger an act that feels natural... a 'no-brainer'. And that zone is exactly what the alien hand jacks into. The visually recognizable objects, along with the information of what to do with them, are hijacked by the hand, without the intentionality of the owner of the said hand. It's in line with what Darian Leader claims: the hands move, they act upon the world. They don't stay still.

Metzinger questions if this 'no-brainer' condition is limited to the alien hand, or if it can extend to the whole body. His proposition becomes curious as some objects put you in a dialectical position upon/immediate to their arrival.

Absolute positions: door, camera, gun, tank...
Soft positions: carpet, microphone, ball, bicycle...

“Tools determine whether goals anticipated mentally can be realised. In that sense tools are never just what they actually are. Rather, they represent the potential of realizing intellectually anticipated goals, which is to say, they represent ideas as real possibilities,” write Damerow and Lefèvre, in *Tools of Science*. “It is man who first makes himself double so as to be a universal for a universal,” writes Hegel, somewhere, I’m told.

There is an accumulation of knowledge of the ways our bodies exist in relation to things around us; as well as, manipulation of the environment and the body through design and feedback. “The use of tools primarily serves the purpose for which they were produced. But... more can always be learned than the knowledge invested in its invention.” Each time a tool is utilized, the user performs a reflection on the experience. The knowledge extracted from this reflection soon after finds itself in the shared realm of social knowledge. Damerow and Lefèvre locate this transfer of information in the representation and abstraction achieved via numbers, letters, terms and techniques. It is in this process that reflection becomes an operator of exponential growth of human agency.

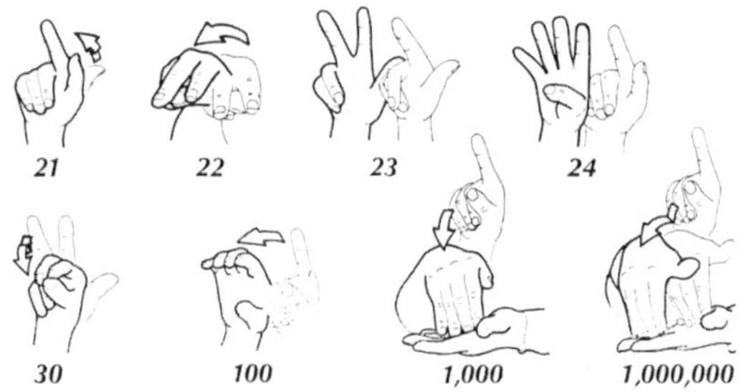

Cognitive Laborer and How to Disown a Limb:

Matteo Pasquinelli talks about a certain recognition or appreciation of the basic cognitive capacities of the normative human, at the onset of artificial intelligence technologies. Comparing the traditionally deemed ‘unskilled’ or purely ‘mechanical’ truck driver with the self-driving car, he sets off to write an extensive book on how AI is fundamentally shaped by labor and social relations.

“Let us return to the contested project of the self-driving car. What kind of work does a driver perform? And to what extent can AI automate such an activity? With a considerable degree of approximation and hazard, a self- driving vehicle is designed to imitate all the micro-decisions that a driver makes on a busy road. Its artificial neural networks 'learn' the correlations between the visual perception of the environment and the mechanical control of the vehicle (steering, accelerating, braking) together with ethical decisions to be taken in a few milliseconds in case of danger. Driving requires high cognitive skills that cannot be left to improvisation, but also rapid problem-solving that is possible only thanks to habit and training that are not completely conscious. Driving remains essentially a social and cooperative activity, which follows both codified rules (with legal constraints) and spontaneous ones, including a tacit cultural code which is different in each locality. It is deemed difficult to encode such a complex activity, and even the entrepreneur Elon Musk has admitted, after not a few fatal accidents of Tesla cars, that 'generalized self-driving is a hard problem'."

Similar to the locomotive driver, the truck driver interacts with a bunch of handles and buttons to pilot the vehicle. Hypothetically though, the truck driver might experience a higher sense of extension of the phenomenal self-image, as their foot is always at the gas pedal and the movement of their foot engages in a continuous channeling of intention. Possibly forgetting the existence of their foot altogether in the act. Driving is best done half-consciously: many good drivers drive out of habit. They'll read the road unconsciously, and will perform rapid decisions reflexively. It is the conscious drivers that often are regarded as bad drivers; all their active processing gets in the way of a smooth drive.

How consciousness stands in the way of its own smooth operation is especially evident when we are learning how to use a tool or a vehicle for the first time. One has to figure out the spatial composition of the situation, by being fully aware of their surroundings and their body's position in it, while at the same time juggling the demand for attention each body part makes. In his *Cronopios and Famas* book, J. Cortázor gives instructions to the simplest of cognitive acts:

"To climb a staircase one begins by lifting that part of the body located below and to the right, usually encased in leather or deerskin, and which, with a few exceptions, fits exactly on the stair. Said part set down on the first step (to abbreviate we shall call it 'the foot'), one draws up the equivalent part on the left side (also called 'foot' but not to be confused with 'the foot' cited above), and lifting this other part to the level of 'the foot,' makes it continue along until it is set in place on the second step, at which point the foot will rest, and 'the foot' will rest on the first. (The first steps are always the most difficult, until you acquire the necessary coordination. The coincidence of names between the foot and 'the foot' makes the explanation more difficult. Be especially careful not to raise, at the same time, the foot and 'the foot'.)"

The distance that Cortázor places between the act of climbing that I already know how to do, and the one that's described above, also leads me to recall the foot that I have just here, right in front of me, and compare with "the foot" that climbs the stairs above. Cortázor offers a route from one to the other so laden with words that a foot becomes disowned/disembodied.

Howling Man and the Night Sky:

A man howls at the other side of the tram station. Howls and howls. Shouts some German words and howls some more. I look for a second, but then I don't want to disturb. I listen from afar.

Later in the tram I am reminded of an old memory. Once in a lifetime event: a completely clear night sky. I was lying on some beach sunbed along with some friends. The moon was not around, city lights were away; and the whole sky was in front of me: the stars and the galaxies all visible... A view I can best describe as a laptop background. Although domesticated and tamed on LCD screens, in the wild with no obstructions, the clear night sky is dangerous. And that night, it produced the most unanticipated vertigo—I remember feeling like I could fall to the sky. Just detach from my sunbed, and fall towards the sky. It felt bizarre as the reasons that kept me grounded became feeble, the up became the down, and I wondered "why wouldn't I fall to the sky?"

The howling man had a similar effect. Why wouldn't I fall into howling? I can see how and why I would. Less and less clear as to how and why I don't.

Unlanguage:

In his peculiar fiction book *Unlanguage* that resembles language learning workbooks, Micheal Cisco imagines a suicide victim that attempts to cross back into the living by learning the undead 'unlanguage of unknowing'. As we accompany him into the classroom, we find out that in this perpetually unstable language, "students should not assume that abnormal spelling or punctuation, when present, are always typological errors."

His classmates' bodies rot and disintegrate, and in the meantime, the student expresses his desires to be a character. However, he can only describe his experience as "being alone, minus something". He exists firmly within the language, in that abstracted state. This we realize because "the verb 'to be' is the only verb in unlanguage." I.e., "the English phrase, 'I say' would be represented in unlanguage by terms meaning 'I am a saying'. The negative, 'I do not say' would be represented by terms meaning 'not I am a saying'."

It is around this time that the fictitious premise of unlanguage starts to crack, and we are confronted by words themselves peeping: "these words and the visions they house describe a life that isn't possible, and can only poison you with futile yearning for what you can't be." "We nouns are dynamisms of some kind of sectioned off from itself, from each-other now and trapped in distinctive baskets of sound or line." The words—to be precise; English words—start to speak to us: "your mind is mummified. There is no life in the people who populate your imagination, your memory, your undead fantasies". It is no longer the fictitious unlanguage Cisco speaks about, but the one we use, when he writes: "In language, no events are unwitnessed! The language itself sees you."

Thankfully, Cisco also offers a way out of this trap, out of this prison. He prompts us to imagine his unlanguage and thus teaches it to us in the meantime. This is a worthwhile attempt because we are told that "the truly advanced student will learn to listen to the unlanguage itself, and to give it the form it most 'desires'". He instructs us: "Do not use your judgement. Allow the characters to take shape randomly. The result will be the aleatory scrawl of the true diviner."

Surprise, Surprise:

There are different levels of intensity with which we look at -perceive- the world. A range of engagement that oscillates between what one can call 'the amnesiac glance' and 'the skeptical glare'. Within the amnesiac-glance-mode, one goes about their day smoothly. Things are as one knows it: remaining fairly within habits, easily recognizable patterns, in "alles in Ordnung". Nothing surprising. It's a mode within which we exist in auto-mode. Did we lock the door? We forget, but we must have. There is not much doubt. Our eyes graze the surface, we don't need to bend down, look past, or squint. We already know, we can switch on that light without even looking.

However, it is in the moment when we reach for the toilet paper and not find it there that another mode of perception and consciousness appears. A surprise. Maybe we forgot to restock the paper, or maybe moved houses and now it hangs on the other side of the toilet in this new bathroom. We remember, we retrace our steps. This is the skeptical-glare-mode, where we become aware of what we see, where we are, the arrangement of things around us, and our position within this arrangement. It gets the brain churning; we have encountered a 'new' and it calls for conscious engagement. Not necessarily a situation that requires assistance of adrenaline, but however small, a surprise makes us pay attention. Either we squint or open our eyes wide, we become aware of our current condition, negotiate our physical or mental positioning, look for patterns, and try to make sense of our experience.

Unlike the smooth operation of 'the amnesiac glance', 'the skeptical glare' is laborious, slow, and hard. But it is precisely what the sci-fi writer and ex-marine-biologist Peter Watts identifies as the actual conscious experience. At a time when scientists still can't convincingly delineate the evolutionary benefits of consciousness precisely because of its non-efficient nature, it does seem to offer some sort of pleasure and excitement to the beholder of the experience. But as with all things, in moderation. People who are exposed to long term threatening conditions may leave the torture chamber unable to close their widened eyes.

In the torture chamber, one gets overwhelmed as things as they know them are no more. Everything can pose danger: all has to be immediately perceived and evaluated. Danger needs to be assessed, so that one can defend against it. This is the only agency possible in this room. Prediction is vital. But things are too new; a lifetime of habits crumble, patterns become unrecognizable, and to top that, there is no time. The high alertness, when sustained for too long, becomes destructive. It corrupts all the effort and time one has spent to make sense of the world, and one's place in it. A process which we rely on completely, because we make causal relations and extract meanings. This is our practice; our method and habit. Can one find peace in a situation where processes of meaning-making are continuously interrupted?

Similar to what Darian Leader identifies in the motor capacities of a baby's hand, which reflexively clasps, we reflexively hold on to our habits, formulated values and judgements. "If clasping is initially a reflex action, releasing is not, and must be learnt over the course of at least the first year." In severely contingent situations similar to the one present in the torture chamber, can one learn to hold themselves together by letting go a bit? Delay judgement, endure uncertainty, and wait?

Oddly, then, 'home' presents itself as the dialectic counterpart to the highly conscious experience of the previous volatile example. It's interesting to think that generation of a sense of 'belonging to' depends on familiarity, and thus, to the lack of consciousness. Considering that it doesn't necessarily need to be the place we sleep in, 'home' can generally be understood as a place that offers stability. A place in which we know where and how things go, about the limits to what the space can hold; we know what to expect. It is a place where we don't need to actively perceive, evaluate and predict things. It's a place for relaxing, place of minimal conscious engagement that makes us feel like we belong to it, precisely because we know it.

But thankfully, the surprise can find us on any level or scale. The smallest trigger can snap us out of an automated existence, may it be the sink clogging or a pimple erupting overnight. The range, though, can extend itself to include devastating earthquakes, wars, or grievous accidents.

Position of the Eye:

Many experiments on the malleability of the phenomenal self-image comes with a single exception. Despite achieving the whole-body analog of the rubber-hand illusion, where the subject felt the mannequin in front of him to be his own body; Metzinger informs that vision remains where it is. In a series of experiments in 2005, Metzinger and his colleagues manage to successfully fool their subjects into believing that a virtual body is their own. A 'jump' into the fake virtual body occurs, where many subjects feel like they are seeing their body from behind. However, although their sense of position changes, their eyes remain in their own body.

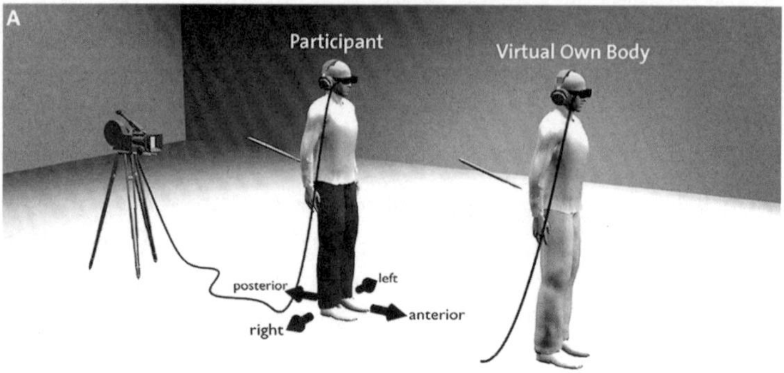

Creating a whole-body analog of the rubber-hand illusion; Participant (darker trousers) sees through a head-mounted display Participant's own virtual body (lighter-colored trousers), standing 2 m in front of Participant and being stroked synchronously at the Participant's back. Illustration by M. Boyer, Recommendations for Good Scientific Practice and the Consumers of VR-Technology, 2016

This prompts Metzinger to make a distinction between bodily self and seeing self—what he calls the 'embedding principle': "The bodily self is phenomenally represented as inhabiting a volume in space, whereas the seeing self is an extensionless point—namely, the center of projection for our visuospatial perspective, the geometrical origin of our perspectival visual model of reality. Normally this point of origin (behind the eyes, as if a little person were looking out of them as one looks out a window) is within the volume defined by the felt bodily self. Yet, as our experiments demonstrated, seeing and bodily self can be separated, and the fundamental sense of selfhood is found at the location of the visual body representation."

If our visual body representation can be relocated outside of our bodies, and our phenomenal self-models can be extended to a point where a tool becomes part of a limb, then the desire for an anchor (a point of origin) becomes understandable. A point of contact to a mass or a surface of our choice, so we don't completely detach, and float adrift.

There is a style of meditation where one starts from the tip of their head, and traces their whole body till the tip of their toes, mentally placing and feeling each body part. Strangely difficult, as the mind doesn't seem to want to engage with this process consciously; it identifies and/or reclaims the visual body representation (phenomenal self-model) and thereby the sense of selfhood. And when this mapping is anchored around the seeing self (right behind the eyes as the origin), it can offer centering as a calming method. Especially when one feels extended beyond their means—worrying over things outside their control, their reach of agency—this method can behave as a re-anchoring.

The trap is that all pre-existing knowledge leads to certain new knowledge that metabolizes back into the system, a process that changes & mutates the model, and on each iteration a new model is born. A supposed causal link can be retroactively argued for, but it cannot be preemptively delineated. It not only lies outside the processing capacity of what was there before, it is also unquestionably disturbed by surprise. Once disenchanted, we escape the lure of history, no longer able to talk about innate qualities to a deterministic end. Control and agency are challenged at each step.

It is unclear why vision does not move, why it stays centered right behind the eyes. It is also unclear how a shrug can be ingrained. Sometimes, we feel like our body is over there, but we are still looking from here.

AN AGENTIAL REALIST ONTOLOGY
Karen Barad

Reality is bigger than us.
–Ian Hacking, *Representing and Intervening*

I think the world is precisely what gets lost in doctrines
of representation and scientific objectivity.
–Donna Haraway, *The Promises of Monsters*

Representationalism takes the notion of separation as foundational. It separates the world into the ontologically disjunct domains of words and things, leaving itself with the dilemma of their linkage such that knowledge is possible. If words are untethered from the material world, how do representations gain a foothold? If we no longer believe that the world is teeming with inherent resemblances whose signatures are inscribed on the face of the world, things already emblazoned with signs, words lying in wait like so many pebbles of sand on a beach there to be discovered, but rather that the knowing subject is enmeshed in a thick web of representations such that the mind cannot see its way to objects that are now forever out of reach and all that is visible is the sticky problem of humanity's own captivity within language, then it becomes apparent that representationalism is a prisoner of the problematic metaphysics it postulates. Like the frustrated would-be runner in Zeno's paradox, representationalism never seems to get any closer to solving the problem it poses because it is caught in the impossibility of stepping outward from its metaphysical starting place. What is needed is a new starting place.

Excerpt from *Meeting the Universe Halfway: Quantum Physics and the Entanglement of Matter and Meaning* (2007), pages 137 – 153, reprinted here with permission from the author & Duke University Press.

The postulation of individually determinate entities with inherent properties is the hallmark of atomistic metaphysics. Atomism hails from Democritus.[1] According to Democritus, the properties of all things derive from the properties of the smallest unit-atoms (the "uncuttable" or "inseparable"). Liberal social theories and scientific theories alike owe much to the idea that the world is composed of individuals with separately attributable properties. An entangled web of scientific, social, ethical, and political practices, and our understanding of them, hinges on the various differential instantiations of this presupposition. Much hangs in the balance in contesting its seeming inevitability.

Niels Bohr won the Nobel Prize for his quantum model of the atom, which marks the beginning of his seminal contributions to the development of the quantum theory. Crucially, however, in a stunning reversal of his intellectual forefather's schema, Bohr rejects the atomistic metaphysics that takes "things" as ontologically basic entities. For Bohr, things do not have inherently determinate boundaries or properties, and words do not have inherently determinate meanings. Bohr also calls into question the related Cartesian belief in the inherent distinction between subject and object, and knower and known. Indeed, Bohr's philosophy-physics poses a radical challenge not only to Newtonian physics but also to Cartesian epistemology and its representationalist triadic structure of words, knowers, and things.[2]

It might be said that the epistemological framework that Bohr develops rejects both the transparency of language and the transparency of measurement; however, even more fundamentally, it rejects the presupposition that language and measurement perform mediating functions. Language does not represent states of affairs, and measurements do not represent measurement-independent states of being. Bohr develops his epistemological framework without giving in to the despair of nihilism or the dizziness of relativism. With brilliance and finesse, Bohr finds a way to hold on to the possibility of objective knowledge as the grand structures of Newtonian physics and representationalism begin to crumble.

1 Atomism is said to have originated with Leucippus and was further elaborated by Democritus, a devotee of democracy, who also explored democracy's anthropological and ethical implications. Democritus's atomic theory is often identified as the most mature pre-Socratic philosophy, directly influencing Plato and Epicurus, who transmitted it into the early modern period. Atomic theory is said to form the cornerstone of modern science.

2 See chapter 3.

Bohr's break with Newton, Descartes, and Democritus is based not in "mere idle philosophical reflection" but on new empirical findings in the domain of atomic physics that came to light during the first quarter of the twentieth century. Bohr's struggle to provide a theoretical understanding of these findings resulted in his radical proposal that an entirely new epistemological framework is required. Unfortunately Bohr does not explore the crucial ontological dimensions of his insights but rather focuses on their epistemological import. I have mined his writings for his implicit ontological views (see chapter 3) and here elaborate on them in the development of an agential realist ontology.

In this section, I present a brief overview of important aspects of Bohr's account and move on to an explication of an agential realist ontology. This relational ontology is the basis for my posthumanist performative account of material bodies (both human and nonhuman). This account refuses the representationalist fixation on words and things and the problematic of the nature of their relationship, advocating instead a *relationality between specific material (re)configurings of the world through which boundaries, properties, and meanings are differentially enacted* (i.e., discursive practices, in my posthumanist sense) *and specific material phenomena* (i.e., differentiating patterns of mattering).[3] This causal relationship between the apparatuses of bodily production and the phenomena produced is one of agential intra-action.[4] The details follow.

According to Bohr, *theoretical concepts* (e.g., position and momentum) are not ideational in character but rather *specific physical arrangements*.[5] For example, the notion of position cannot be presumed to be a well-defined abstract concept; nor can it be presumed to be an individually determinate attribute of independently existing objects. Rather, position has meaning only when an apparatus with

3 Some philosophers may see the move I am making as naturalistic. I have labeled this approach "posthumanist" rather than "naturalist" because the considerations that go by the former term are also interested in troubling the nature-culture distinction (though it is important to recognize that there are many different posthumanisms), whereas "naturalism" (which also designates multiple stances) generally holds the nature-culture dualism in place. Rather than presuming an inherent distinction between nature and culture, I am interested in accounting for how this distinction is made and remade. See Rouse 2004a on understanding agential realism as a form of naturalism that does not abide by the nature-culture dualism.

4 Causality from an agential realist perspective is discussed hereafter.

5 Bohr argues on the basis of this single crucial insight, together with the empirical finding of an inherent discontinuity in measurement intra-actions, that one must reject the presumed inherent separability of observer and observed, knower and known. See chapter 3.

an appropriate set of fixed parts is used. And furthermore, any measurement of position using this apparatus cannot be attributed to some abstract, independently existing object but rather is a property of the *phenomenon*—the inseparability of the object and the measuring agencies. Similarly, momentum is meaningful only as a material arrangement involving a specific set of movable parts. Hence the indeterminacy of simultaneous position and momentum measurements is a straightforward matter of the material exclusion of position and momentum arrangements (one requiring fixed parts, and the complementary arrangement requiring those same parts to be movable).

As I argued in chapter 3, the primary ontological unit is not independent objects with inherent boundaries and properties but rather *phenomena*. In my agential realist elaboration, phenomena do not merely mark the epistemological inseparability of observer and observed, or the results of measurements; rather, *phenomena are the ontological inseparability entanglement of intra-acting "agencies."* That is, phenomena are ontologically primitive relations—relations without preexisting relata.[6] The notion of intra-action (in contrast to the usual "interaction," which presumes the prior existence of independent entities or relata) represents a profound conceptual shift. It is through specific agential intra-actions that the boundaries and properties of the components of phenomena become determinate and that particular concepts (that is, particular material articulations of the world) become meaningful. Intra-actions include the larger material arrangement (i.e., set of material practices) that effects an *agential* cut between "subject" and "object" (in contrast to the more familiar Cartesian cut which takes this distinction for granted). That is, the agential cut enacts a resolution within the phenomenon of the inherent ontological (and semantic) indeterminacy. In other words, relata do not preexist relations; rather, relata-within-phenomena emerge through specific intra-actions. Crucially, then, intra-actions enact *agential separability*-the condition of *exteriority-within-phenomena*. The notion of agential separability is of fundamental importance, for in the absence of a classical ontological condition of exteriority

6 That is, relations are not secondarily derived from independently existing relata; rather, the mutual ontological dependence of relata—the relation—is the ontological primitive. As discussed later, relata only exist within phenomena as a result of specific intra-actions (i.e., there are no independent relata, only relata-within-relations). The term "intra-action" signifies the mutual constitution of relata within phenomena (in contrast to "interaction," which assumes the prior existence of distinct entities). In particular, the different agencies remain entangled.

between observer and observed, it provides an alternative ontological condition for the possibility of objectivity. Moreover, the agential cut enacts a causal structure among components of a phenomenon in the marking of the "measuring agencies" ("effect") by the "measured object" ("cause"). It is in this sense that the measurement can be said to express particular facts about that which is measured; that is, the measurement is a causal intra-action and not "any old playing around."[7] Hence *the notion of intra-action constitutes a reworking of the traditional notion of causality.*[9]

In my further elaboration of this agential realist ontology, I argue that phenomena are not the mere result of laboratory exercises engineered by human subjects; rather, *phenomena are differential patterns of mattering* ("diffraction patterns") produced through complex agential intra-actions of multiple material-discursive practices or apparatuses of bodily production, where *apparatuses are not mere observing instruments but boundary-drawing practices-specific material (re)configurings of the world-which come to matter.* These causal intra-actions need not involve humans. Indeed, it is through such practices that the differential boundaries between humans and nonhumans, culture and nature, science and the social, are constituted.[9]

Phenomena are constitutive of reality. Reality is composed not of things-in-themselves or things-behind-phenomena but of things-in-phenomena.[10] The world is a dynamic process of intra-activity and materialization in the enactment of determinate causal structures with determinate boundaries, properties, meanings, and

7 The phrase "any old playing around" is a reference to Schrödinger's expressed concern that the notion of measurement retain some sensible meaning even if the properties expressed do not preexist the measurement process: "In general, a variable has no definite value before I measure it; then measuring it does not mean ascertaining the value that it has. But then what does it mean? There must still be some criterion as to whether a measurement is true or false, a method is good or bad, accurate, or inaccurate—whether it deserves the name of measurement process at all. Any old playing around with an indicating instrument in the vicinity of another body, whereby at any old time one then takes a reading, can hardly be called a measurement of this body" (Schrödinger 1935, 158). See chapter 7 for more details.

8 In *How Scientific Practices Matter*, Rouse takes up my notion of causal intra-actions and situates Bohr's analysis in terms of philosophical debates about causation.

9 For a discussion of the nature of the role of the experimenter and other human subjects, see hereafter.

10 Because phenomena constitute the ontologically smallest unit, it makes no sense to talk about independently existing things as somehow behind or as the causes of phenomena. In a sense, there are no noumena, only phenomena. Agential realist phenomena are not Kantian phenomena or the phenomenologist's phenomena.

patterns of marks on bodies. This ongoing flow of agency through which part of the world makes itself differentially intelligible to another part of the world and through which causal structures are stabilized and destabilized does not take place in space and time but happens in the making of spacetime itself. It is through specific agential intra-actions that a differential sense of being is enacted in the ongoing ebb and flow of agency.[11] That is, it is through specific intra-actions that phenomena come to matter-in both senses of the word.

The world is an open process of mattering through which mattering itself acquires meaning and form through the realization of different agential possibilities. Temporality and spatiality emerge in this processual historicity. Relations of exteriority, connectivity, and exclusion are reconfigured. The changing topologies of the world entail an ongoing reworking of the notion of dynamics itself. Dynamics are a matter not merely of properties changing in time but of what matters in the ongoing materializing of different space-time topologies. The world is intra-activity in its differential mattering.

In summary, the primary ontological units are not "things" but phenomena-dynamic topological reconfigurings/entanglements/relationalities/(re)articulations of the world. And the primary semantic units are not "words" but material-discursive practices through which (ontic and semantic) boundaries are constituted. This dynamism *is* agency. Agency is not an attribute but the ongoing reconfigurings of the world. The universe is agential intra-activity in its becoming.

In what follows, I provide a detailed explication of this agential realist ontology. I begin with a detailed examination of the nature of the apparatus, including two significant analytical shifts that are important emendations to Bohr's formulation: (1) a shift from linguistic representations to discursive practices; and (2) a shift from apparatuses as static prefab laboratory setups to an understanding of apparatuses as material-discursive practices through which the

11 The metaphor of "flow" is used here to conjure the idea of agency as an enactment, as opposed to its more usual conception as a property of entities (most commonly humans). On the other hand, it is also misleading in that this metaphor connotes a sense of fluidity that is premised on dynamics that change continuously in time, whereas intra-actions take place not in space and time (as if they were containers or markers of what already is) but in the making/marking of spacetime. Crucially, this profound shift goes beyond the usual geometrical concepts of dynamics to a reconceptualization of "dynamics" itself where continuity is not assumed and topological shifts (e.g., changes in continuity, boundedness, etc.) may be of great significance.

very distinction between the social and the scientific, nature and culture, is constituted.

THE NATURE OF AN APPARATUS

The opportunity to know the apparatus better.... That is an integral part of knowing how to create phenomena.
–Ian Hacking, *Representing and Intervening*

What is an apparatus? Is it the set of instruments needed to perform an experiment? Is it a meditating device that allows the object world to give us a sign of its nature? Is it a prosthetic extension of our sensing abilities? Shall we understand an apparatus in terms of Kantian grids of intelligibility? Aristotelian schemata? Heideggerian background practices? Althusserian apparatuses? In Foucault's sense of discursive practices or dispositif(apparatus)? In Butler's sense of the performative? As Latour's inscription or translation devices? Or as Haraway's apparatuses of bodily production? Bohr's notion of an apparatus is unique among these theorizations, and yet there are some interesting reverberations among these possibilities worth exploring. Reverberating at different frequencies, these differing lines of thought can productively be read through one another for the patterns of resonance and dissonance that illuminate new possibilities for understanding and for being.

Since apparatuses play such a crucial, indeed constitutive, role, it is imperative that we understand their precise nature. In this chapter I argue that apparatuses are not mere instruments or devices that can be deployed as neutral probes of the natural world, or determining structures of a social nature, but neither are they merely laboratory instruments or social forces that function in a performative mode. Apparatuses are not merely about us. And they are not merely assemblages that include nonhumans as well as humans. Rather, apparatuses are specific material reconfigurings of the world that do not merely emerge in time but iteratively reconfigure space-time matter as part of the ongoing dynamism of becoming.

THE BOUNDARY OF AN APPARATUS

Bohr specifies certain specific criteria for apparatuses. According to Bohr, apparatuses are macroscopic material arrangements through which particular concepts are given definition, to the exclusion of others, and through which particular phenomena with particular determinate physical properties are produced. The far-reaching conclusion of Bohr's proto-performative analysis is that the apparatus plays a much more active and intimate role in experimental practices than classical physics recognizes. Apparatuses are not passive observing instruments; on the contrary, they are productive of (and part of) phenomena. Yet despite the centrality of the apparatus to Bohr's analysis, he never fully articulates its nature.

Questioning the basis of the Newtonian tradition, Bohr refuses to take for granted the delineation of the "object" and the "agencies of observation" and makes the constitution of this "inside" boundary the centerpiece of his analysis. In particular, he emphasizes that the cut delineating the object from the agencies of observation is enacted rather than inherent. On the other hand, Bohr does seem to help himself to the "outside" boundary of the apparatus. That is, while focusing on the lack of an inherent distinction between measuring instrument and measured object, Bohr does not directly address the question of where the apparatus "ends." Is the outside boundary of the apparatus coincident with the visual terminus of the instrumentation? What if an infrared interface (i.e., a wireless connection) exists between the measuring instrument and a computer that collects the data? Does the apparatus include the computer? Is the printer attached to the computer part of the apparatus? Is the paper that is fed into the printer? Is the person who feeds in the paper? How about the person who reads the marks on the paper? Or the scientists and technicians who design, build, and run the experiment? How about the community of scientists who judge the significance of the experiment and indicate their support or lack of support for future funding? What precisely constitutes the limits of the apparatus that gives meaning to certain concepts at the exclusion of others?

One of the questions, perhaps *the* question, that Bohr finds most pressing in his investigation of measurement practices is how it is possible to secure the conditions for the possibility of objectivity given that "subjective elements" such as human concepts play a productive (though not determining) role in the outcome of measurements. In other words, what is at stake for him in the challenge posed by quantum physics is nothing less than how we can account for the fact that science works. Crucial to Bohr's analysis of the subject-object distinction is his insistence that concepts are materially embodied in the apparatus. In particular, Bohr insists that only concepts defined by their specific embodiment as part of the material arrangement-which includes instrumentation (e.g., photographic plates, pointers, or digital readout devices) that marks definite values of the specifically defined properties and can be read by a human observer-are meaningful. That is, the larger material arrangement enacts a cut that resolves the inherent ontic-semantic indeterminacy through which the "subject" and the "object" emerge. Apparatuses are the conditions of possibility for determinate boundaries and properties of objects and meanings of embodied concepts within the phenomenon. Indeed, this embodiment of concepts as part of the apparatus is ultimately what secures the possibility of objective knowledge, as defined in terms of Bohr's epistemic criteria of reproducibility and communicability.[12] One pronounced limitation of Bohr's account, then, is that the human is thereby cemented into the very foundations of the quantum theory and the far-reaching philosophical implications of his proto-performative account of scientific practices. Observation and communication, the contingencies of visibility and invisibility, of concepts and utterances, are crucial to this formulation: man isn't merely the measure of all things; man's finitude is implicated in the very conditions of possibility of measurability and determinability. It is as if in the desire to compensate for the shortcomings of classical mechanics—which erroneously jettisons the observer from the scene of observation—Bohr overshoots his mark and places the human not merely back in the picture where she or he belongs, but at the center of all that is.

Furthermore, Bohr's conception of the apparatus does not accurately take account of the complexities of experimental practice, despite his apparent intentions to do just that. Quite atypical of the writings of theoretical physicists, Bohr's papers include detailed drawings of measuring instruments. His attentiveness to the details of the apparatus makes perfect sense given his insistence that the concepts used to describe phenomena are not ideations but specific material arrangements: for Bohr, word and world are tied to each other. Nonetheless Bohr treats the apparatus itself as an ideal measuring device that springs full blown from the head of Zeus, operates itself or at most requires the pushing of a few buttons to produce

12 See chapter 3 for more details.

results, requires no tinkering, no maintenance, no muss, no fuss. Its constitutionality remains constant—no rearrangements, no alterations, no adjustments. It is frozen at a moment in time, denied its historicity and mutability. In an important sense, then, Bohr's apparatus is hermetically sealed off from any and all "outside" influences. The scientist is a liberal humanist subject who is merely there to choose an appropriate apparatus for the investigation and note the results. Once the apparatus is in place, the scientist stands back and watches what happens.

In short, Bohr mistakes the apparatus for a mere laboratory setup. Magically, the scientific instrumentation works correctly without intervention, reducing the role of the experimenter to a mere recorder of the objective marks displayed by the instrumentation.[13] For all of Bohr's insistence on thinking realistically about apparatuses, refusing to contemplate them as idealized forms, he artificially cuts "the apparatus" off from all the activities that enable experimental practice to work. As Hacking notes:

Most experiments don't work most of the time. To ignore this fact is to forget what experimentation is doing.

To experiment is to create, produce, refine and stabilize phenomena. . . . But phenomena are hard to produce in any stable way. That is why I spoke of creating and not merely discovering phenomena. That is a long hard task.

Or rather there are endless different tasks. There is designing an experiment that might work. There is learning how to make the experiment work. But perhaps the real knack is getting to know when the experiment is working. That is one reason why observation, in the philosophy-of-science usage of the term, plays a relatively small role in experimental science. Noting and reporting readings of dials—Oxford philosophy's picture of experiment—is nothing. Another kind of observation is what counts: the uncanny ability to pick out what is odd, wrong, instructive or distorted in the antics of one's equipment. The experimenter is not the "observer" of traditional philosophy of science, but rather the alert and observant person. Only when one has got the equipment running right is one in a position to make and record observations. That is a picnic. (1983, 230)

13 This can perhaps be likened to the mistaken belief that cookbook laboratory exercises teach students about experimentation. They don't. Rather, at best they are heuristic devices for learning theoretical notions, but even this is questionable (see Barad 2000b).

Bohr's figuration of the apparatus makes it inoperable in practice.[14]

The liberal humanist conception of the subject and the taken-for-granted static and bounded apparatus that are embodied in Bohr's theoretical apparatus get in the way of his efforts to provide a deeper understanding of the nature of scientific practices and ultimately cut short the profound ontological implications of his ideas. In taking for granted an intrinsic outside boundary of the apparatus, which incorporates human concepts within its bounds while ejecting the observer to the outside, Bohr reifies the selfsame "subjective elements" he sets out to tame, ignoring the dynamism of discursive practices and the co-constitution of subjects along with objects. Bohr seems to have forgotten his own lesson that cuts are part of the phenomena they help produce. What is needed is a posthumanist understanding of the role of the apparatus and of the human and the relationship between them.

TOWARD AN AGENTIAL REALIST UNDERSTANDING OF APPARATUSES

One task that stands before us is to further elaborate Bohr's immensely important insights while removing the less savory anthropocentric elements, including his dependence on the notion of human concepts and laboratory setups. As we have seen, it is also important to take into account the dynamic and complex nature of scientific practices. In recent years, critical social theorists have offered sophisticated accounts of the practices by which meanings, boundaries, and bodies are produced. The problem is that these accounts are invested and enmeshed in a host of anthropocentric assumptions as well. For example, Judith Butler's performative account of mattering thinks the matter of materiality and signification together in their indissolubility; however, Butler's concern is limited to the production of human bodies (and only certain aspects of their production, at that), and her theorization of materialization is parasitic on Foucault's notions of regulatory power and discursive practices, which are limited to the domain of human social practices. Furthermore, for both Butler and Foucault, agency belongs only to

14 The recent shift in focus from the study of scientific knowledge to the study of scientific practice has contributed to the acknowledgment of the complexity of experimentation. See, for example, Hacking 1982, 1983; Galison 1987, 1997; Pickering 1984.

the human domain, and neither addresses the nature of technoscientific practices and their profoundly productive effects on human bodies, as well as the ways in which these practices are deeply implicated in what constitutes the human, and more generally the workings of power. That is, both accounts honor the nature-culture binary (to different degrees), thereby deferring a thoroughgoing genealogy of its production.[15] Crucially, Butler's and Foucault's theories fail to provide an adequate account of the relationship between discursive practices and material phenomena, leading one to wonder ifBohr's insights into the embodied nature of concepts might fruitfully intervene in this regard. What is needed is a posthumanist performative account of the material-discursive practices of mattering (including those that get labeled "scientific" and those that get labeled "social").

In what follows, I diffractively read the insights of Bohr, Foucault, Butler, and other important theorists through one another in an effort, to advance such an account. This will entail important reworkings of the notions of materiality, discursive practices, agency, and causality, among others. My agential realist elaboration of apparatuses entails the following significant developments beyond Bohr's formulation: (1) apparatuses are specific material-discursive practices (they are not merely laboratory setups that embody human concepts and take measurements); (2) apparatuses produce differences that matter-they are boundary-making practices that are formative of matter and meaning, productive of, and part of, the phenomena produced; (3) apparatuses are material configurations/dynamic reconfigurings of the world; (4) apparatuses are themselves phenomena (constituted and dynamically reconstituted as part of the ongoing intra-activity of the world); (5) apparatuses have no intrinsic boundaries but are open-ended practices; and (6) apparatuses are not located in the world but are material configurations or reconfigurings of the world that re(con)figure spatiality and temporality as well as (the traditional notion of) dynamics (i.e., they do not exist as static structures, nor do they merely unfold or evolve in space and time).

15 See chapters 1 and 5 for more details.

MATTERING: A POSTHUMANIST PERFORMATIVE ACCOUNT OF MATERIAL-DISCURSIVE PRACTICES

Discourse is not a synonym for language.[16] Discourse does not refer to linguistic or signifying systems, grammars, speech acts, or conversations. To think of discourse as mere spoken or written words forming descriptive, statements is to enact the mistake of representationalist thinking. Discourse is not what is said; it is that which constrains and enables what can be said. Discursive practices define what counts as meaningful statements. Statements are not the mere utterances of the originating consciousness of a unified subject; rather, statements and subjects emerge from a field of possibilities. This field of possibilities is not static or singular but rather is a dynamic and contingent multiplicity.

According to Foucault, discursive practices are the local sociohistorical material conditions that enable and constrain disciplinary knowledge practices such as speaking, writing, thinking, calculating, measuring, filtering, and concentrating. Discursive practices produce, rather than merely describe, the subjects and objects of knowledge practices. In Foucault's account, these conditions are immanent and historical rather than transcendental or phenomenological. That is, they are not conditions in the sense of ahistorical, universal, abstract laws defining the possibilities of experience (Kant), but actual historically and culturally specific social conditions.

Foucault's account of discursive practices has some provocative resonances (and some fruitful dissonances) with Bohr's account of apparatuses and the role they play in the material production of bodies and meanings. For Bohr, apparatuses are particular physical arrangements that give meaning to certain concepts to the exclusion of others; they are the local physical conditions that enable and constrain knowledge practices such as conceptualizing and measuring; they are productive of (and part of) the phenomena produced; they enact a local cut that produces "objects" of particular knowledge practices within the particular phenomena produced. On the basis of his profound insight that "concepts" (which are actual physical arrangements) and "things" do not have determinate boundaries, properties, or meanings apart from their mutual intra-actions, Bohr offers a new

16 I am concerned here with the Foucauldian notion of discourse (discursive practices), not formalist and empirical approaches stemming from Anglo-American linguistics, sociolinguistics, and sociology.

epistemological framework that calls into question the dualisms of object-subject, knower-known, nature-culture, and word-world.

Bohr's insight that concepts are not ideational but rather actual physical arrangements is clearly an insistence on the materiality of meaning making that goes beyond what is usually meant by the frequently heard contemporary refrain that writing and talking are material practices. Nor is Bohr merely claiming that discourse is "supported" or "sustained" by material practices, as Foucault seems to suggest (though the nature of this support is not specified), or that nondiscursive (background) practices determine discursive practices, as some existential-pragmatic philosophers purport.[17] Rather, Bohr's point entails a much more intimate relationship between concepts and materiality, matter and meaning.

The shift from linguistic concepts to discursive practices provides the possibility of freeing Bohr's account from its reliance on human concept and the static nature of apparatuses in one move. At the same time, however, the notion of discursive practices must be appropriately reconceptualized to take account of their intrinsically material nature (and Bohr's insights are helpful here). The basic idea is to understand that it is not merely the case that human concepts are embodied in apparatuses, but rather that apparatuses *are* discursive practices, where the latter are understood as specific material reconfigurings through which "objects" and "subjects" are produced. I will offer such an elaboration in what follows. This shift will include a proposed posthumanist understanding of discursive practices and the role of the human, as well as some other important considerations. The agential realist ontology provides a basis for the necessary elaborations.

In my agential realist elaboration of Bohr's account, *apparatuses are the material conditions of possibility and impossibility of mattering*; they enact what matters and what is excluded from mattering. Apparatuses enact agential cuts that produce determinate boundaries and properties of "entities" within phenomena, where

17 Foucault makes a distinction between "discursive" and "nondiscursive" practices, where the latter category seems to be reduced to social institutional practices: "The term 'institution' is generally applied to every kind of more-or-less constrained behaviour, everything which functions in a society as a system of constraint and which isn't utterance, in short, all the field of the non-discursive social, is an institution" (Foucault 1980, 197–98; italics mine). This specific social science demarcation is not particularly illuminating in the case of agential realism's posthumanist account, which is not limited to the realm of the social. In fact, it makes no sense to speak of the "nondiscursive," per se, given my posthumanist conception of discursive practices as boundary reconfigurings that are inherently material and need no material support.

"phenomena" are the ontological inseparability of agentially intra-acting components. That is, agential cuts are at once ontic and semantic. It is only through specific agential intra-actions that the boundaries and properties of "components" of phenomena become determinate and that particular articulations become meaningful. In the absence of specific agential intra-actions, these ontic-semantic boundaries are indeterminate. In short, the apparatus specifies an agential cut that enacts a resolution (within the phenomenon) of the semantic, as well as ontic, indeterminacy. *Hence apparatuses are boundary-making practices.*

Now, as Bohr and Foucault would no doubt agree, meaning should not be understood as a property of individual words or groups of words. Meaning is neither intralinguistically conferred nor simply extralinguistically referenced. Meaning is made possible through specific material practices. Semantic contentfulness is achieved not through the thoughts or performances of individual agents but through particular discursive practices. However, the common belief that discursive practices and meanings are peculiarly human phenomena won't do. If discursive practices are boundary-making practices in an ontic (as well as semantic) sense, then the practices by which the human and the nonhuman are differentially constituted cannot rely on a notion of discursive practices that helps itself to a prior notion of the human. What is needed, then, is a posthumanist understanding of discursive practices.

In an agential realist account, *discursive practices are specific material (re)configurings of the world through which the determination of boundaries, properties, and meanings is differentially enacted.*[18] That is, *discursive practices are ongoing agential intra-actions of the world* through which specific determinacies (along with complementary indeterminacies) are enacted within the phenomena produced. Importantly, *discursive practices are causal intra-actions*-they enact causal structures through which some components (the "effects") of the phenomenon are marked by other components (the "causes") in their differential articulation. Meaning is not a property of individual words or groups of words but an ongoing performance of the world in its differential dance of intelligibility and unintelligibility.

18 Which is not to suggest that meaning is ever fixed once and for all. Discursive practices entail ongoing contestations and re(con)figurings. (Nor is it to suggest that all meanings are set are once—recall Bohr's point about complementary conditions for meaning making.) Ambiguities always exist.

In its causal intra-activity, part of the world becomes determinately bounded and propertied in its emergent intelligibility to another part of the world, while lively matterings, possibilities, and impossibilities are reconfigured. Discursive practices are boundary-making practices that have no finality in the ongoing dynamics of agential intra-activity.

In traditional humanist accounts, intelligibility requires an intellective agent (that to which something is intelligible), and intellection is framed as a specifically human capacity. But in my agential realist account, intelligibility is an ontological performance of the world in its ongoing articulation. It is not a human-dependent characteristic but a feature of the world in its differential becoming. The world articulates itself differently.

Furthermore, knowing does not require intellection in the humanist sense, either. Rather, knowing is a matter of differential responsiveness (as performatively articulated and accountable) to what matters. As such, agential realism goes beyond both humanist and antihumanist accounts of the knowing subject as well as recent insights concerning the knower as a prosthetically enhanced human. Knowing is not about seeing from above or outside or even seeing from a prosthetically enhanced human body. Knowing is a matter of intra-acting. Knowing entails specific practices through which the world is differentially articulated and accounted for. In some instances, "nonhumans" (even beings without brains) emerge as partaking in the world's active engagement in practices of knowing.[19] Knowing entails differential responsiveness and accountability as part of a network of performances. Knowing is not a bounded or closed practice but an ongoing performance of the world.

Discursive practices are not speech acts, linguistic representations, or even linguistic performances, bearing some unspecified relationship to material practices. Discursive practices are not anthropomorphic placeholders for the projected agency of individual subjects, culture, or language. Indeed, they are not human-based practices. On the contrary, agential realism's posthumanist account of discursive practices does not fix the boundary between human and nonhuman before the analysis ever gets off the ground, but rather allows for the possibility of a genealogical analysis of the material-discursive emergence of the human. Human bodies and

19 See the description of the brittle star in chapter 8.

human subjects do not preexist as such; nor are they mere end products. Humans are neither pure cause nor pure effect but part of the world in its open-ended becoming.

Just as there are no words with determinate meanings lying in wait as so many candidates for an appropriate representational moment, neither are there things with determinate boundaries and properties whirling aimlessly in the void, bereft of agency, historicity, or meaning, which are only to be bestowed from the outside, as when the agency of Man pronounces the name that attaches to specific beings in the making of word-thing pairs. "Things" don't preexist; they are agentially enacted and become determinately bounded and propertied within phenomena. Outside of particular agential intra-actions, "words" and "things" are indeterminate. Matter is therefore not to be understood as a property of things but, like discursive practices, must be understood in more dynamic and productive terms-in terms of intra-activity.

In *Bodies That Matter*, Judith Butler gives a thorough accounting of the failures of social constructivist accounts of the body that circulate in feminist theory and challenges feminists to return to the notion of matter. But by this "return" she does not advocate reclaiming the precritical view that would position matter as that which is prior to discourse. She argues that any such attempt to ground feminist claims about sexual difference in such a prediscursive substance is doomed to beach itself on that very shore: matter, she explains, is already "fully sedimented with discourses on sex and sexuality that prefigure and constrain the uses to which that term can be put" (Butler 1993, 29). Instead Butler proposes that we understand matter as a "process of materialization that stabilizes over time to produce the effect of boundary, fixity, and surface we call matter" (9). She explains that her claim that "matter is always materialized has . . . to be thought in relation to the productive and, indeed, materializing effects of regulatory practices in the Foucaultian sense" (9-10).[20]

Butler's reconceptualization of matter as a process of materialization brings to the fore the importance of recognizing matter in its historicity and directly challenges representationalism's construal of matter as a passive and blank slate awaiting the active inscription of culture whereby the relationship between materiality and

20 For more details, see chapter 1.

discourse is figured as one of absolute exteriority. Butler's account emphasizes the following important points. Matter, like meaning, is not an individually articulated or static entity. It is not little bits of nature, or a blank slate, surface, or site passively awaiting signification; nor is it an uncontested ground for scientific, feminist, or economic theories. Matter is not immutable or passive. Nor is it a fixed support, location, referent, or source of sustainability for discourse. It does not require the mark of an external force like culture or history to complete it. Matter is always already an ongoing historicity.

Unfortunately, however, Butler's theory ultimately reinscribes matter as a passive product of discursive practices rather than as an active agent participating in the very process of materialization.[21] This deficiency is symptomatic of an incomplete assessment of the causal factors of materialization and an incomplete reworking of "causality" in understanding the nature of discursive practices (and material phenomena) in their productivity. Furthermore, Butler's theory of materiality is limited to an account of the materialization of human bodies or, more accurately, to the construction of the contours of the human body. Moreover, as her reading of materiality in terms of Foucauldian regulatory practices makes clear, the processes that matter for her are only human social practices (thereby reinscribing the very nature-culture dichotomy she wishes to contest). Agential realism provides an understanding of materialization that goes beyond the anthropocentric limitations of Butler's theory. Significantly, it recognizes matter's dynamism.

In an agential realist account, matter does not refer to a fixed substance; rather, *matter is substance in its intra-active becoming-not a thing but a doing, a congealing of agency. Matter is a stabilizing and destabilizing process of iterative intra-activity.* Phenomena—the smallest material units (relational "atoms")—come to matter through this process of ongoing intra-activity. "Matter" does not refer to an inherent, fixed property of abstract, independently existing objects; rather, *"matter" refers to phenomena in their ongoing materialization.*

Matter is not simply "a kind of citationality" (Butler 1993, IS), the surface effect of human bodies, or the end product of linguistic or discursive acts. Matter is not a linguistic construction but a discursive production in the posthumanist sense that discursive practices are themselves material (re)configurings of the world

21 See discussion in chapter 1.

through which the determination of boundaries, properties, and meanings is differentially enacted. That is, discursive practices as boundary-making practices are fully implicated in the dynamics of intra-activity through which phenomena come to matter. The dynamics of intra-activity entail matter as an active "agent" in its ongoing materialization. Or rather, *matter is a dynamic intra-active becoming that is implicated and enfolded in its iterative becoming. Matter(ing) is a dynamic articulation/configuration of the world.* In other words, materiality is discursive (i.e., material phenomena are inseparable from the apparatuses of bodily production; matter emerges out of, and includes as part of its being, the ongoing reconfiguring of boundaries), just as discursive practices are always already material (i.e., they are ongoing material [re]configurings of the world). Discursive practices and material phenomena do not stand in a relationship of externality to each other; rather, *the material and the discursive are mutually implicated in the dynamics of intra-activity.* The relationship between the material and the discursive is one of mutual entailment. Neither discursive practices nor material phenomena are ontologically or epistemologically prior. Neither can be explained in terms of the other. Neither is reducible to the other. Neither has privileged status in determining the other. Neither is articulated or articulable in the absence of the other; matter and meaning are mutually articulated.

Material constraints and exclusions and matter's historiality and agency (including, for example, the material dimensions of regulatory practices) are important factors in the process of materialization.[22] Material conditions matter, not because they "support" particular discourses that are the actual generative factors in the formation of bodies, but because matter comes to matter through the iterative intra-activity of the world in its becoming. The point is not merely that there are important material factors in addition to discursive ones; rather, the issue is the conjoined material-discursive nature of constraints, conditions, and practices. The fact that material and discursive constraints and exclusions are intertwined points to the limited validity of analyses that attempt to determine individual effects of material or discursive factors.[23]

22 Derrida's term "historiality" rather than "historicity" is perhaps more appropriate. See the following section on re(con)figuring spacetimematter.

23 For further discussion, see chapters 5 and 6.

Agential realism's conceptualization of materiality makes it possible to take account of material constraints and conditions once again without reinscribing traditional empiricist assumptions concerning the transparent or immediate givenness of the world and without falling into the analytical stalemate that simply calls for recognition of our mediated access to the world and then rests its case. The ubiquitous pronouncements that experience or the material world is "mediated" have offered precious little guidance about how to proceed. The notion of mediation has for too long stood in the way of a more thoroughgoing accounting of the empirical world. The reconceptualization of materiality offered here makes it possible to take the empirical world seriously once again, but this time with the understanding that the objective referent is phenomena, not the seeming "immediately given-ness" of the object world.[24]

All bodies, not merely "human" bodies, come to matter through the world's iterative intra-activity—its performativity. This is true not only of the surface or contours of the body but also of the body in the fullness of its physicality, including the very "atoms" of its being. Bodies are not objects with inherent boundaries and properties; they are material-discursive phenomena. "Human" bodies are not inherently different from "nonhuman" ones. What constitutes the human (and the nonhuman) is not a fixed or pregiven notion, but neither is it a free-floating ideality. What is at issue is not some ill- defined process by which human-based linguistic practices (materially supported in some unspecified way) manage to produce substantive bodies or bodily substances, but rather the dynamics of intra-activity in its materiality: material apparatuses produce material phenomena

24 Referentiality is reconceptualized. Words do not refer to things as in representationalist accounts. Rather, words and things are intra-actively coarticulated through specific patterns of practice. The world differentially articulates itself. To speak of the "objective referent" is not to presume to put into relation disparate entities (i.e., words and things or words and phenomena) but to acknowledge the relationality of the intra-active coarticulation of matter-discourse. Rouse (2002) calls this an expressivist account: "The material-inferential relations between patterns of talk and particular practical interactions (including experimental practices) both articulate the meaning or content of what is said and express what is going on in the practical interactions. In the case of experimental science, I will argue, this expressive role of scientific discourse is not something external to the phenomena investigated, but is a constitutive component of the phenomenon itself" (269). Kirby (1997) puts it this way: "What I am trying to conjure here is some 'sense' that word and flesh are utterly implicated, not because 'flesh' is actually a word that mediates the fact of what is being referred to, but because the entity of a word, the identity of a sign, the system of language, and the domain of culture—none of these are autonomously enclosed upon themselves. Rather they are all emergent within a force field of differentiations that has no exteriority in any final sense" (126–27).

through specific causal intra-actions, where "material" is always already material-discursive—*that is what it means to matter*. Theories that focus exclusively on the materialization of human bodies miss the crucial point that the very practices by which the differential boundaries of the human and the nonhuman are drawn are always already implicated in particular materializations. The differential constitution of the human (nonhuman) is always accompanied by particular exclusions and always open to contestation. This is a result of the nondeterministic causal nature of agential intra-actions, a crucial point that I take up in the section on agency and causality hereafter.

FROM = TO ≈ AND ≘
A CONSIDERATION ON JUSTICE AFTER QUANTUM MECHANICS
Raqs Media Collective

An allegorical figure of justice had lost its head in a museum basement. We brought it out and placed it in a vitrine, lit it, looked at it carefully, and invited others to offer it their consideration as well.

A figure of 'headless justice' might lead one to surmise that this is what happens to 'justice' when it loses its capacity to reason. What if one presumed a different possibility, of a 'headless' kind of reason? Could 'headless' justice have gained something in return for losing her head?

Is equal to?

Ordinarily, when the idea of justice is invoked in any conversation, it presupposes a relationship between an action and a consequence, or an equation of contending claims - to attention, to significance, or to resource.

How, ordinarily, does justice 'reason'?

Every time the idea of justice is invoked, it presupposes a relationship between an action and a consequence, or an equation between more than one claim (to attention, to significance, to resources). Justice is indexed by the appropriateness, or commensurability, of that relationship or equation.

This concept may be expressed by the arithmetical symbol of *is equal to* = sign. The expressions, entities, claims, and quantities that stand on either side of *is equal to* are taken to be of equivalence. They need not be identical, but ideally, they need to be 'equal' in a 'counting' sort of way. Justice is indexed by the appropriateness, or commensurability, of that relationship or equation.

Over millennia, ever since Hammurabi's code specified the 'eye for an eye' principle, we have been accustomed to an understanding of justice which is quantitative, based on metrics, and in part, on the = sign. If a person has blinded another in one eye, then, the blinding man's eye must also be taken away. Not two eyes, but one eye, because the taking away of two eyes of a one-eye-blinder would be 'unjust'. At the same time, it means that there is no way by which any idea of mercy, or compassion, can be entertained. One cannot say, 'all right, let's not blind the blinder, not even in one eye'.

Perhaps this mode of thinking has even influenced what we consider to be a 'Newtonian' framework - of 'every action' having 'an *equal* and *opposite* reaction'. This presumes that we can measure the action and the reaction, and see that they are 'equal, and opposite'. This 'Newtonian' principle has become the bedrock of the vengeful mode of justice. Harm must be answered by another harm, only then can things be restored to a 'baseline' state that they existed in, before being 'disturbed' by the first harm that set the spiral of consequential harm into motion. Again, this presumes an ability to measure harm, and to divide harm into discrete packets that can be weighed against each other, as they spin and spiral.

Justice = Revenge?

The narrative of popular cinema too tends to be deeply committed to this modus operandi. This is how things turn out: first, a peaceful world is established, with nice, peaceful people going about their business. Then a terrible wrong committed by a vengeful oppressor, unkind power for its own selfish ends, disturbs the tranquil equilibrium of this peaceful world. The institutions of justice fail to deliver a sense of recompense. A hero, perhaps in hiding, nursing past injustice, in stoic solitude, is called upon to do what he can. He acts, at first reluctantly, or hesitantly, or ineptly and then heroically, to exact an exact revenge. Or maybe he goes overboard, just a little, as if he were only paying harm back to compensate for harm, with interest.

In any case, crime must be recompensed by punishment, because there is no other way, in this universe, of dealing with harm.

That's the Newtonian Mechanics of Justice as 'revenge'. Predicated on a measure of injury, it fuels not just the heroics of mainstream cinema, but also the rhetoric of politics, representative or not. First, this grammar of politics has to recognise harm, then it has to identify a perpetrator, then it has to promise to punish the perpetrator, and finally, it has to amplify what it undertakes in order to propagate the violence of the harm, and the justification of revenge through which justice is 'shown' to be done.

Is equilibrium a state of justice?

To combat revenge justice, take the = sign, introduce a wave form to queer its straight edges, the double tilde sign ≈ an orthographic innovation designed to denote approximation. A single tilde sign ~ and layering it above an = sign also produces an interesting result ≌ the sign for congruence or alignment.

Approximation, congruence, alignments are estimates of relationality, they are signs that remain open to discussion, they occupy, not distinct points, but a spectrum, a gradient; they are not claims to exact measurements of identity that can be used to put an end to a conversation.

A glimmer in contention?

When a Śūdra has sex with a guarded or unguarded woman of a twice-born class - he loses a limb and all his possessions, if she was unguarded. If she was guarded, a Śūdra loses everything. A Vaiśya is imprisoned for a year and all his property is confiscated; and a Kṣatriya is fined 1,000 and his head is shaved using urine. If a Vaiśya or a Kṣatriya has sex with an unguarded Brahmin woman, the Vaiśya is fined 500 and the Kṣatriya 1,000. If any of these two have sex with a guarded Brahmin woman, he should be punished in the same way as a Śūdra or he should be burnt with a straw fire. A Brahmin who has forced sex with a guarded Brahmin woman should be fined 1,000; for sex with a willing partner, he should be fined 500.
–The quotation is a translation of sections 374 – 364 (pertaining to 'adultery') in Chapter 8 of the *Manusmriti*, translated by Patrick Olivelle, as 'Manu's Code of Law', (page 187).

A granular operation of codification of life, of the rules of conduct along a graded hierarchy governed by caste and patriarchal rules, is deeply enmeshed in the conception of bodies, obedience, and punishment in the subcontinent of India and, in a large measure, draws from the approximately 2000-year-old legal text called *Manusmriti*. Kings governed with counsel aided by this text. The hold of the priestly castes on the interpretation of virtue, and vice, drew from this. The extensive delineation of rules over women's body is at the core of this code.

The Manusmriti is no longer 'law' in India, and the text is publicly set on fire by anti-caste activists every year, keeping alive a nearly 100-year-old defiant practice. But despite being rendered irrelevant by the Constitution of the Republic in India in 1950, the code remains alive in the way in which the question of *daṇḍa*, i.e. the scale of the punishment, gets framed. The rules of how much *daṇḍa*, and to whom is reserved the severest *daṇḍa*, has followed the graded inequality of the caste hierarchy, and fallen harshest on people from the castes placed lower, and on women, and on indigenous tribes.

The code forbids the allowance of a speaking subject for the majority of the populace, and keeps it limited to a small coterie of governing castes.

Films, songs, literature, and various treatises have at times tried to undercut the weight of the ritual and psychic possession of Manu's classification and sensibility of conduct, but a Newtonian reflex—of avenging as a natural force against a perceived wrong—brings them in proximity to the realm of punishment offered by Manu. The overturning of the caste and gender rules governing the Speaking Subject continues to get thwarted in the aligning of deviation back into the same and legible moral and social order.

An experimental attempt is made by Pa Ranjith in his film 'Natchathiram Nagargiradh' ('Wandering Star', 2022), to dare another, a different, a not-yet-taken, path. He complicates the question of how justice can look if the speaking subject arrives from a very different phenomenology of reading actions, injury, and punishment. Who can then speak, and then who can speak *to* all and *for* all, is a question that Pa Ranjith explores and asks us to deliberate on.

'Wandering Star' is set within the dynamics of an amateur theatre group that is developing a play on the theme of love and how it is bound by prohibitions and proscriptions emanating from caste, religion, and gender norms. A male actor, whose controlling misogyny and homophobia is made evident through the film, aggressively accosts and harasses a female actor during a party while under the influence of excessive intoxication. She is not the only person he aggresses against at the party; his hostile attentions extend to two senior members of the group who try to restrain him from drinking more than he can hold, and to queer members of the group. The other members of the diverse theatre group include, besides heterosexual men and women, a gay male couple, a lesbian couple, an inter-racial couple, an interfaith couple, and a man in a relationship with a transwoman. They all express sympathy and solidarity with the woman. They decide, at the urging of the head of the group, to ask the molester to leave the group. This decision is communicated to him on the morning after the party in which the incidents took place, and when he comes to apologise to everyone.

Ranjith's filmography has repeatedly dissected the persistence of caste norms in Tamil society despite the state's long history of anti-caste and social justice politics. In this film itself we see the female protagonist's unforgiving disdain and despair at the condescension of her upper-caste (Brahmin) ex-boyfriend about how her caste identity inflects her personality, and also the dynamics where a dominant intermediate caste family is unable to deal with the prospect of their son marrying the 'lower' caste woman he is in love with. Ranjith alerts the viewer even to micro-aggressions that cannot make sense expect for when seen through the prism of caste. His narrative enables characters to move beyond a restoration of inter-personal equilibria, towards a rewriting of the transcripts of caste in a language that takes them outside the realm of private and personal injury to something much more foundational: the calculus of offense, injury, and justice itself.

While the man who had aggressed at the party, after expressing remorse and apologising to the group, as well as to the woman he had accosted, is about to leave, the woman argues against ostracization, and that the group set up a demand and a condition that he can stay should he address and change his aggressive, homophobic, misogynistic behaviour. The group re-thinks it's 'collective conscience' and the aggressor is asked to stay and prove that he can change. What this work of change will look like is for us to collectively postulate and speculate on. It is a journey that will have to draw from different histories of jurisprudence, moral ecologies, failed or aborted experiments and, most crucially, it returns the question of justice to the granularity of daily life and the conversations within it.

The film frames the woman's stance on a sensibility that emerges from her Dalit sense of self and history, in which there is awareness that to differentiate the crime from its context in the matrix of caste and patriarchy is to not address the crime at all. We could read this as an argument that recognises that violence emerges not only from an individual's action, but also from an overarching structure of normative behaviour that teaches aggression and legitimises the aggressor. In a sense, it's like what the forensics school of analysis would argue: the crime becomes a molecular moment that unzips all kinds of sediments and structures that make possible the very occurrence of the crime. It is within this framework of what justice could mean that the protagonist asks not for punishment but for the acknowledgment of transformation as living practice. A practice that does not sanction or overlook or forgive a perpetrator, but opens out the possibility of a resolve towards self-transformation, thus making the social order unstable and inverted.

How can one measure pain, or humiliation?

Following the problem of observation in quantum mechanics, one could say that the very act of observing pain leads to a change in the circumstances in which pain is perceived. In quantum mechanics, physicists tend to approximate and conjecture, and can never abstract themselves away from the problem they are engaged with. The observer is part of the experimental reality that she is observing.

In the 'Newtonian' model of justice, justice is external to what it judges, which is why it is able to objectively 'measure' harm or injury and settle on a 'just' recompense. The quantum model does not allow for this 'objective' Archimedean point of external reference. It includes the judge and the judged, the witnesses and the advocates, in an entangled meshwork. This means that there is no 'all seeing' eye, located in a 'head' that stands, outside, and above, the 'scene of the crime'.

What kind of conversation do we have in mind?

Imagine a conversation between bones, between scrambled heaps of disinterred human remains from mass graves, some arranged to form nearly complete, notionally discrete, skeletons, for the benefit of a war photographer, to illustrate casualties, long after the fact of war, on a battlefield. The photograph, *Scene at Secundrabagh* is dated 1858, and staged for and taken by the photographer Felice Beato.

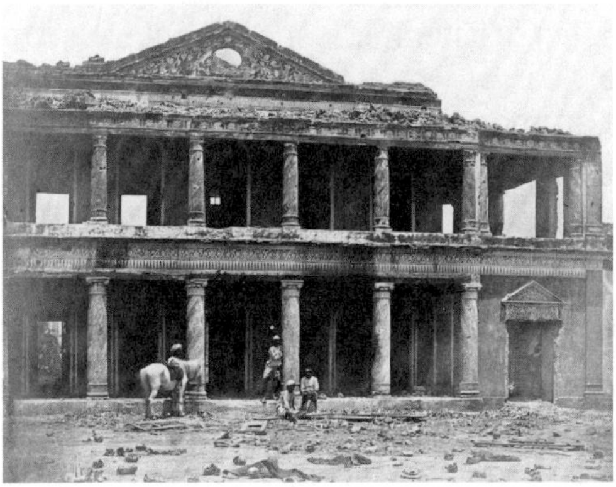

Albumen silver print, Felice Beato, 1858, Wikimedia Commons

An imagined conversation between bones, a cadaverous cacophony -

...if each of the two hundred and six bones of the adult human body could speak, they would all sing the body's praises. Ossuaries would be operas.

The sternum, or the breastbone would testify to its owner's pride, the ribs would act as a sentimental chorus, singing an elegy to the fluttering heart-bird, the fibula, tibia and femur, would drum up anthems to strength and vigour, the tarsals and metatarsals, carpals and metacarpals, radius and ulna would praise poise and dexterity, the coccyx would hit a base complaining note, the frontal bone of the cranium would worry. Bones of every function and description would whisper, scream, speak in tongues and measured tones, laugh, cry, sing in tune, off-key and off-kilter. Only the lonesome hyoid, the unarticulated bone of the tongue, might choose silence in jest at the excess of cadaverous cacophony. The hyoid would hold its tongue, knowing that every life is deserving of only as much noise, or silence, as every other life.

The noise, speech, song, and silence of bones multiply the attributes of what it means to be human in a way that defies the imperative of measurement. They enable comparisons between different states of life and death, but disable exact computations of the value of life and death. If one were to pay attention and practice listening to the arias, duets, choruses, and recitatives from an ossuary-opera, one might find reason to revise ideas of how the living and the dead may receive justice. Despite the impossibility of measure, each life is '*as deserving of noise, or silence, as every other life*'.

To enable this act of listening, justice might have to 'lose' her head, and find her ears close to the ground, so as to listen better to matters that may be as intimate as they are subterranean.

Notes

'Opera of the bones', in *The Surface of Each Day is a Different Planet*, 2009, Film (38 minutes) https://works.raqsmediacollective.net/index.php/2009/10/26/the-surface-of-each-day-is-a-different-planet/3815/

LATENT FIELDS OF HYPER-HISTORY: QUANTUM & METAPHYSICS FOR AFRI-INDIGENOUS CHARACTERS IN RECENT SCIENCE FICTION

Bogosi Sekhukhuni

[Latent fields can be understood as a psycho-spacial shape describing at least some of the possible motilities that allow anti-life forms to inhabit memories of all experiences preceding and exceeding the experiences of the present moment. Anti-life forms are how the dead and the unborn and the actions of the dead and the unborn occupy the present moment, and this is because there cannot be a present without the actions of the dead and the unborn. The experience of the present already betrays its own expectation. The present of the unborn is the same present of the dead. If anything, let us continue with the understanding that when we experience the present we're moving into environments that are the petrified remains/entrails of the actions of the dead and the unborn. The distant fate of all that is right now is to become a detail on the contour of hyper-history. Latent fields come out of an exploration of grand scientific fictions. The phrase scientific fiction is not meant to be paradoxical. Instead, it seeks to redefine how both concepts can be evaluated.]

I have enjoyed thinking about some recent characters coming out of North American science fiction TV productions in the late 2010's and early 2020's. The Protagonist from Christopher Nolan's *Tenet*, the season two story arc surrounding Commander Michael Burnham and Dr Gabriel Burnham in CBS Studio's *STAR TREK: Discovery*, Sergeant Roberta Draper's story arc in season 2 of the

James S.A Corey book and TV series *The Expanse,* and Client Zero from my video work *Simunye Systems Orientation Video*. The characters I'm curious about are portrayed by Afri-descendent and Indigenous performers. Each character navigates environments of dominant histories, working with and at odds with the spacetimes they occupy.

To stay with the spaciotemporal tensions facing racialised characters, I want to touch on Michelle M. Wright's notion of "epiphenomenal time" in her critique of dominant linear progress narratives. In *Physics of Blackness: Beyond The Middle Passage Epistemology*, Wright challenges what she describes as dominating schools of thinking centering the Black Atlantic experience and contrasts this observation by offering definitions of Blackness that highlight the multidimensionality of Black selfhood, are contoured by historical processes, and acknowledge multiple polarities of ongoing trans-geographic connectivity. Wright exercises her positions through an analysis of the experiences of characters drawn from diasporic Black speculative fiction.

In the spirit of her book, what follows is a basic discussion on and through quantum theories of uncertainty. The aim of which is to try to extend or take seriously the relationships of metaphors Wright creates between her idea of epiphenomenal spacetime and the physical theories that play a role in its formulation. Latent fields is a concept I introduce to further support this attempted extension in order to begin to satisfy an intuition about the emergent spacetime logics that contribute to epiphenominal spacetime.

A most enduring theme in science fiction (SF) involves spacetime-affecting technologies that disrupt the dominant temporalities in the worlds of such characters. This trope offers SF practitioners a kind of toolkit for the reinventing of histories and collective identities. In a community dominated still by a Eurocentric imagination, what follows are some initial thoughts on Black and Indigenous characters in recent SF.

The kind of thinking through spacetime temporalities by SF practitioners that are of interest for now, directly and indirectly, complicate what Wright calls a Newtonian "linear progress narrative". Wright points out that linear progress narratives are regularly instrumentalised by both diasporic and non-diasporic civilization projects. Tracking these projects across space and time feels important but for now, the Western cultural project of Enlightenment is a useful example of a civilization project that self-mythicises grand narratives. Foundational to such narratives is a belief in the 'naturally' occurring progression of history. Myths of progress are often recounted as idealised, forward marches towards some irradiated, ascension machine, coming out of/from a darker, wilder, and less evolved past.

Myths of progress appear to be fundamental to nationalist projects. In them is a compelling and prevailing feature; that of a grand narrative in which the collective triumphs over an oppressive authority. More intimately nuanced variations of this narrative structure exists across geographies but fundamentally, the origin stories of a nation and the cultures of commemoration produced around these origin stories represent a position that is in opposition to an antagonistic, imposing force. Regardless of the merit of an origin story, as these concepts boil through processes of political, ideological rhetoric and other cultural forces, the historical past is at increasing risk of being flattened and adjusted to rhyme with a given nationalist origin narrative.

For Wright, this tendency is representative of a hegemonic, uncritical reliance on the picture of reality offered by classical mechanics. Wright argues that there is a lasting, seductive power in the ubiquitous logic provided by classical mechanics in daily macro life. We experience many aspects of life in linear ways. Newton's mathematics are fundamental to understanding the way tangible bodies move through atmosphere-rich environments, or calculations of the motions of planetary bodies that are necessary for space navigation.

Epiphenomenal time is a concept Wright offers to stretch the temporal capacity for the present moment, promoting a nonlinear, relativistic positioning in spacetime. It is an outlook or condition that de-centers any singular account or perspective of an event as primary, or 'native'. Instead, epiphenomenal time favors the primacy of an experience of spacetime involving all activities that, from a Newtonian perspective, exist in the 'past' or the 'future'. Wright expands on this position by qualifying some entropic qualities of epiphenomenal time; noting that our experience of time is not consecutive but rather accumulative. What ties events of the past to the present is not a causal connection in the literal sense.

"THE OUT OF AFRICA HYPOTHESIS REINFORCES DOMINANT LINEAR PROGRESS NARRATIVES"

Symptomatic of Wright's progress myths is the immediate negation of histories of thought that aren't rooted in territories where myths of progress are acted out. Amongst the enduring intellectual and cultural products of European enlightenment is the franchised Other. A totalising dump housing essentialist categories describing everything that is outside of Europeanness and/or outside of—in its subverted appropriation by modern thinkers—hegemony. I think that investigations on otherness point to the inability of Wright's linear progress narratives to account for parallel, divergent and recursive epistemological and ontological timelines and realities that persist irrespective of the Enlightenment project.

Forms of connection that Past and Future events have to the Present are produced and managed by the Present. Maybe another way to put it is that at any given point, what exists of recorded historical events is produced in the exact, individual, present moment when participants are motivated to historicize or remember a past event. Acts of recollection involve multiple factors, some of which are important for a high resolution historical event. Levels of resolution impact what kind of historical potential is absorbed by the present. A low resolution inspires unmitigated speculation and an overbearing influence of existing hegemonic states. (Eventually making fertile conditions for the development of cultures of anti-intellectualism, ethno-nationalist propagandas and pseudosciences.)

Epiphenomenal time reminds me of ideas that complicate modern formulations about temporality; from concepts in endogenous Bantu metaphysics to climatic determinist critiques in the environmental humanities. A recent correlate I have come across is Walter Benjamin's meditations through a historical materialist approach to history. In the essay *On the Concept of History,* Benjamin calls for a psycho-spiritually immersive engagement with events in the past by removing a sense of everything that happens after; to summon or imagine a historical context or memory as if it were taking place in your sense of the present moment.

Curiously, Benjamin invokes the shock of trauma that comes from such a psychic displacement as a catalyst for this kind of recollection (perhaps making it possible?). In discussing characters of interest, I might want to think about a function of this sort of traumatic break as a protocol for reversing or reconstructing internal systems of entropy.

Traumatic events have the effect of splintering one's sense of spacetime. To elaborate a bit further, when there is a subscription to a dominant linear progress narrative, it is dysregulated by the imprint of a traumatic event. The imprint preserves this event. The individual must now accept a reality in which they experience both the present and the past simultaneously.

I would like to take a stab at thinking about a model for the present moment, or *changelings,* that organises both past and future events as indeterminate states informing and updating the other. And, possibly describing at least some of the ways that Wright's conception of epiphenominal time [or spacetime] might be organised. Firstly, Wright's approaches and registers of spacetime are conceptually possible, in part, because of the uncertainty principle and the law of wave-particle duality. When it comes to going about the business of material existence, indeterminacy is a fundamental nature and materiality for both an observer's position in spacetime and may be for the structure of spacetime itself. We can understand a position in spacetime to relate to historical resolution, with increasing or decreasing historical resolution correlating with the nearness or distance that an event has to the rememberer in their present.

Indeterminate states of past contexts and future anxieties carry dissimilar densities partly because of their orientation. When we look at spacetime, cosmologically speaking, the orientation of the time in spacetime is singular, a world line glowing with an ever-expanding light cone. This unidirectional flow is not necessary for an Einsteinian understanding of spacetime but is of interest to our conversation on latent fields in that it might be evidence of an emergent quality, perhaps a higher dimensional structure, underlying or supporting spacetime.

In invoking a past or a future-directed spacetime, we influence how the present now is culturally, cognitively and perhaps physically defined. The influence I'm referring to is the impact of the interlacing of past contexts, present events, and future anxieties. For Wright, past and future cease to be relevant descriptors. In a Latent fields framework, each form of present exists independently and collectively as *changelings*. Changelings are shaped by the ways that each one builds resolutions of its different parts.

The fuzziness of material realities of past and future spacetime events diffuses the spacetime we experience as the present constant—its motility is expressive of the complexity of exchanged histories, both imagined and real. The density of the fuzziness is determined by collective and individual memory, providing various levels of resolution enforcing the boundaries of the present.

This explains our sense of the present now as a fluctuating condition. I'm interested in thinking about a bright description projecting Wright's phenomenological framework onto the object of cosmological spacetime. To take this seriously, the epiphenomenal experience of spacetime would have to be universally applicable and at work even through Wright's linear progress narratives. After all, Einsteinian physics does not invalidate the laws of classical mechanics.

By necessity, the present is a latent field because to embody it requires an awareness beyond what the present moment can actually account for. Latent fields might be useful to continue thinking about the kind of condition that is created by epiphenomenal time and spacetime. They are informational fields that describe various degrees of clarity or spaciotemporal resolution. I'm basically offering the idea that epiphenomenal spacetime is not just guided by entropic principles; that there are additional psycho-spaciotemporal structures that potentially support this experience. That, given the historical record of disenfranchised communities globally—and for the purposes of this discussion—in looking particularly at racialised communities of the Global South; an awareness and aptitude for navigating latent fields is readily available for these populations and, hot take: can arguably be understood as a kind of embodied inheritance or consequence of histories of subjectivity.

TIME CRYSTALS AND QUANTUM ERASERS

The Protagonist in the film *Tenet* finds himself undergoing a psychological and material unorganising of selfhood. With the help of Mumbai-based arms dealer Priya, and the agent Neil, the Protagonist uncovers an epiphenomenal experience of time first as a reversed missed bullet shot in the opening scene. The bi-directional flow of trains that frame the closing scene of this sequence emphasises this introduction to a multi-temporal regime. Broadly speaking, the material calibrations I'm referring to are the geopolitical entanglings of emergent temporal warfare technologies and the unfolding nature of the Protagonist's intelligence operation.

Priya informs the Protagonist that in the future, a scientist develops an algorithm with the ability to 'inverse' the entropy of any object (and apparently, the object that is spacetime). Fearing the technological potential for mutually assured destruction scenarios between past and future generations of humans, the scientist resorts to "...splitting the algorithm into nine sections" and dispersing each section into their 'past'. In the audience's timeline, a "cat and mouse jaunt" ensues between the Protagonist and his distastefully brute Eastern European billionaire antagonist.

Psychically, the Protagonist's experience of legal death in the opening scene of the film is itself an inversion. A resetting of the biological clock whose passing is more ambiguous. The Protagonist is forced to embrace the logics available to him.

In Neil's final scene, he shares with the Protagonist that he had "... a future in the past, years ago for me... years from now for you" and it is revealed that Neil, Priya, and all the collaborators working with the Protagonist were under the employment of a future-originating Protagonist operating a "temporal pincer movement". This is described as a warfare tactic where an actor or team "...moves forward through an event", observes the outcomes of the engagement, and then "attacks backwards" having hindsight of how events unfold. A pattern of motion not dissimilar to the electrons or photons, cited by Wright, in John Wheeler's delayed-choice thought experiment, that appear to be able to perceive and adjust for 'future' changes made to their paths.

The audience shares in both these internal and external recoveries as disclosures of memory. Critiques of this popular interpretation of results of experiments based on Wheeler's delayed-choice thought experiments remind us that particles exhibit a dual nature, simultaneously existing as particle-like and wave-like "entities". I think Wright leaves us plenty of room to play with her metaphysical engagement with John Wheeler's delayed-choice experiments. In 1978, Wheeler, esteemed physicist and philosopher of science, shared a proposal for an experiment that investigates the wave-particle dual nature that quanta somehow embody.

Wheeler's experiments are based on the double-slit experiment devised almost 200 years prior by the polymath, Thomas Young. Building on 17th and 18th-century science describing the wave theory of light, in 1801, Young sought to disprove the Newtonian assertion that light acts like a particle. Young demonstrated the wave-like nature of light by imagining an experiment where a light source is beamed at a series of cardboard barriers. The first barrier has one slit, both letting the light through and focusing it as it passes through a second barrier with two slits. The slits each measured under a millimeter thick, and, when Young conducted this experiment, after the light passed through the second screen, it would impact onto a third screen as an interference pattern.

By the mid-20th century, it was thought that a single unit of quantum matter shot at a double-slit screen would pass through one slit or the other. The discrete nature of particle matter gives us this intuition. However, even the use of heavier particles of matter demonstrates the same interference effect. Significant for several reasons, Young's experiment establishes in the Western sciences the basic idea that particles have both wave-like and particle-like qualities.

Wheeler's "delayed-choice" version of the double-slit experiment considers the problem of observation or more specifically, detection. Wheeler wondered about the Earth-bound path of light from a distant quasar, he explains that a photon from this quasar would have to choose which paths through space to take, navigating the gravitational pull of some galaxy cluster or black hole; traveling, on either side as a particle or in both directions as a wave. either observations are possible to detect, depending on the chosen astronomical tools. Inferometers do not detect the exact positions of individual photons, instead, they analyse the collective behavior of a group of photons. Wheeler tried to understand how the choice of observational tools, made from an Earthly perspective millions of years after the photon left its origin, creates the very observation of wave-like or particle-like behavior.

This question encouraged efforts in modern experiments to add detectors with a "delayed choice" element to a double-slit type environment. The 'choice' is the decision of whether or not to measure the path of the particle, the decision being 'delayed' until after the particle has entered the slit or splitter. In 1999, Yoon-Ho Kim, R. Yu, S. P. Kulik, Y. H. Shih, and Marlan O. Scully published findings from an experiment that sent a pair of entangled photons through an environment of instruments combining the delayed-choice and quantum eraser experiments. Variations of the delayed-choice and quantum eraser environments are organised as a complex set of paths, beam splitters, and mirrors to both record and erase information about the paths of a beam and its entangled twin. The Quantum eraser adds another step to the delayed-choice experiment. In a setup where a particle travels along a path that is being measured [occurring after the other ½ of the entangled particle has been detected at the end of its path], the path information of the particle being measured is erased.

Yoon-Ho Kim and colleagues found that their results were still consistent with previous experiments; the appearance of an interference pattern by what is usually referred to as the 'signal' photon was consistent with whether the longer path of its entangled twin, the 'idler' photon, was known, even when the measurement was erased after the 'signal' photon was detected.

It is tempting to think about retrocausality when thinking about this quantum effect. However, what's more interesting for a latent field about the nature of wave-particle duality as demonstrated by a delayed-choice environment is less about a perceived ability to predict the future and more about the role that probability plays to an observer experiencing the unfolding of spacetime. When we look at particles we make them visible to our 4 dimensional material and temporal reality. When we are not focused on a particle, it is distributed across a specific range of possible positions in a state of superposition that probably only feels counterintuitive from a spacetime awareness that sees changelings as firmly bound and distinct.

The payoff at the end of *Tenet* is the revelation of the Protagonist's larger temporal pincer movement. The Protagonist's character development in adapting to the shifting temporal environments he faces is ultimately bound by the plans set in motion, or detected (in keeping with the analogy with Wheeler's delayed-choice thought experiment) by the future Protagonist.

The second season of *Star Trek Discovery* revolves around extreme epiphenomenal spacetime effects that the Red Angel time travel technology has on its users, namely Dr Gabriel Burnham and her daughter Michael Burnham. *Star Trek Discovery* writers imagine a spacetime travelling technology powered by time crystals.

Classical and Quantum Time crystals are an emergent research area in physics. While a mineral-based crystal found in nature is structured by atomic patterns that repeat in all spatial directions, the atomic behavior of a time crystal breaks the principle of time translation symmetry by occupying alternating positions at different points in time, repeatedly. Translation symmetries describe the idea that fundamental laws of physics are consistent across spacetime. Time crystals are atomically arranged to oscillate periodically, exhibiting a ground state that repeats indefinitely, in the dimension of time. Such systems are intrinsic, don't have any energy input and to my understanding, expand our understanding of physical laws of conservation, like entropy.

Dr Burnham is introduced through Michael's deepest childhood trauma as the parent who, along with Michael's father, died at the hands of Klingon soldiers when Michael was a child. We find out that Dr Burnham escaped the attack by travelling 950 years into the future only to discover a galaxy in ruins, devoid of all sentient life.

Dr Burnham became consumed with the future threat. For the past 20 years of Michael's life, Dr Burnham undertook 840 trips across a millennium, attempting to change the past using the Red Angel suit. In her time, Michael Burnham inadvertently ends up contributing to this work by reengineering the Red Angel suit and piloting 7 successive trips, 5 into her recent past and two trips into and from the future.

Powered by time crystal technology, Michael and Dr Gabrielle move with agility in epiphenomenal spacetimes. This is suggested particularly for Dr Burnham. In every attempt she makes to travel back to the present time she is native to, Dr Gabrielle is pulled 'back' 950 years to her original destination. It is as if she was a macro sized atomic member of a non-equilibrium field, existing in multiple places at different periods. Dr Gabrielle finds power in her knowledge of the future extermination of all life and works across time for survival.

Michael Burnham's skill as agent of epiphenomenal spacetime is enhanced through her genetic connection to her mother. It is worth thinking about how this dynamic might contribute to Wright's thinking about "horizontal as well as vertical connections". Furthermore, the new physics of matter introduced by time crystal experiments is interesting to consider in our understanding of the physical concepts that might contribute language to Wright's epiphenomenal spacetime.

In the second season of *The Expanse*, Martian Sergeant Roberta Draper is embroiled in a geo-spatiotemporal conflict across the dominant planetary epistemes of Earth and Mars. Mars is populated by dome-dwelling human colonies and is envisioned as a second

Eden; a celebration of human ingenuity weighted on the fragile dreams of a terraformed atmosphere. This mythology openly echoes popular culture's sympathetic attitude towards Martian exploration and colonisation in the contemporary world.

I AM REQUESTING POLITICAL ASYLUM ON EARTH

Sergeant Draper is a nationalist and is passionate and anxious to serve Mars. Her anxiety is partially informed by the socio-political tensions existing between Earth and Mars. *The Expanse* establishes a complex interplanetary history where Mars is envisioned as "An entire nation dedicated to a common goal, to turn a lifeless rock into a garden". An idealism paired with a militant paranoia about the dominance of Earth's UN government in the solar system, arguably speaking to a desire by Martian territories to transcend their status as a colony of Earth. The Expanse explores intergenerational differences amongst Martians. Sergeant Draper moves through epiphenomenal spacetime in her first trip to Earth as the surviving witness to an armed conflict.

Earth is the past that Sergeant Draper inherited from the early settlers. To me this means that even though Draper is native to Mars, the vertical ties that support their journey back, towards, or up towards Earth, continue classical postcolonial relationships from Earth to Mars. The history of the terraforming of Mars in the solar system cosmology of *The Expanse* feeds a kind of climate determinist grand narrative that for a Martian identity in the world of *The Expanse*, becomes a linear progress narrative fueled by colonisation.

Draper, like the rest of her Martian delegation, is impacted by Earth's gravity. This is an apt metaphor as in a short space of time, Draper uncovers her own native military's complicity in the shootout that resulted in the killing of their crew. She makes an escape and defects to the UN. Draper is interesting in that her nationalism takes her beyond Martian linear progress narratives and to an extent, beyond the limitations of her physiology. Earth's gravitational force is over twice as strong as Mars. The physiological strain on a body acclimated to a Martian atmosphere mirrors Draper's psychological strain; an ideological disillusionment from planetary reorientation. For Draper there is no escaping the weight of her connection to her collective past, She is completely consumed by this past even as she experiences rapid growth in her present constant, or latent field. Like Michael Burnham, it is the full body immersive transportation across spacetime through which Draper serves an evolving political agenda that actively engages epiphenomenal spacetimes.

LATENT FIELDS: ETERNAL INFLATION

Through the collective work on non-Euclidean geometries by 19th century mathematicians Carl Friedrich Gauss and Bernhard Rieman and the half-a-century-later advancements made by Albert Einstein and Hermann Minkowski, a mathematical description of material reality emerged in the 20th century unifying space and time. It is a description that is sufficiently reliable for the infrastructures societies build and exist. Riemannian geometry, a branch of differential geometry, introduced the western scientific world to the possibility of describing, mathematically, surfaces or objects with more dimensions than forward or backwards, and left or right. Riemann built on early 19th century non-euclidean geometry to describe spaces with any number of dimensions, allowing for the first time since Euclid, a mathematical description of the kinds of surfaces that exist in the material world.

Since Reimann, the concept of higher dimensions has been influential to modern, Western arts and cultures. As Reimannian geometry provided a mathematical foundation for Einstein to describe the curvature of spacetime, it also inspired literature, theology and arts exploring alternative worlds. In Reimann's framework and in other non-euclidean mathematics of the 19th and early 20th centuries allowing for higher dimensional space; the existence of dimensions of reality hidden from our shared, 3-dimensional reality.

Simunye Systems Orientation Video is set in KWAAL WORLD, a fictional version of Earth. KWAAL WORLD mirrors our own in many ways except for a history of a more advanced technological progress, one where Magic developed hand in hand with various sciences important to the collective infrastructures of the societies in KWAAL WORLD. The protagonist referred to as Client Zero, is a Johannesburg native in a post-apartheid South African political landscape characterized by conspiracy and dystopian analogies. Client Zero is the first gene donor of a biotechnology company that offers amongst other services, insurance plans catered to a patented analysis of the account bearer's genetic profile. Upon submitting a sample at a pop-up station, Client Zero experiences a psychic attack and is pursued by android technicians operating the pop-up station. In the epilogue, a year has passed and Client Zero appears in a landscape of antediluvian ruins. There is a suggestion that a dimensional portal technology is accessible here.

The existence of KWAAL WORLD is supported by several cosmological theories in which a multiverse of universes is possible. A combination of frameworks and tools developed in higher dimensional mathematics allow cosmologists to describe space and time as a unified phenomenon interacting with mass and energy. These concepts describe a curvature of spacetime, measure distances, positions and relationships between it and bodies of mass and light. Frameworks such as the metric tensor formulated by Hermann Minkowski in 1908, are fundamental to current cosmological models of the universe. Minkowski advanced on the connection between space and time in Einstein's special relativity by creating a framework unifying the 3 dimensions of space with an additional dimension of time in one 4-dimensional model. The Minkowski metric provided a geometric solution to Einstein's early work on relativity and variations of it have been used ever since.

The Standard cosmological model - also known as the Lambda Cold Dark Matter Model- predicts a mostly flat universe and makes use of a variation of the metric known as the Friedman-Lemaître-Roberston-Walker metric. In cosmology Lambda is a number known as the cosmological constant, representing the amount of energy in empty space while cold dark matter refers to dark matter that moves slower relative to the speed of light. Its slow movement lets it clump together and form the structures for galaxies and clusters. The FLRW metric, describes the homogeneous, isotropic and expanding nature of the distribution of matter in the universe. It also predicts a flat shaped universe. Lambda CDM makes use of the FLRW metric to describe the large scale structures and evolution of the universe.

The Lambda CDM model borrows concepts from Inflation theory, in which the early universe experiences a period of rapid expansion at an exponential rate doubling in size every fraction of a second, propelled outward, by a hypothetical field called the inflaton. During inflation, quantum fluctuations spread across this cosmic medium, creating variations of energy density. Such density perturbations are the seeds of the cosmic structures that emerge after the big bang. Following this, in the concept of eternal inflation, the inflation process did not end after the big bang but continues in some areas of space, creating pockets or bubbles of expanding spacetime. In a multiverse scenario these bubbles might grow large enough to form galaxies and universes unlike our own, each evolving out of unique initial conditions.

For all four characters described, the awareness of an epiphenomenal spacetime perspective was unveiled by the imprint of a traumatic event. This imprint is transformative for the existing ways of being each character embodies, in the way that it forces the individual to engage with a multitude of simultaneous, often incompatible spacetimes. It should be noted that the trauma experiences each character undergoes is thought about to the extent of the influence that those imprints have on their temporalities. The Protagonist survives legal death and enters an adjusted relationship with citizenry. Captain Michael Burnham witnesses the resurrection of a deceased parent whose own engagement with epiphenomenal spacetime has contrasting rhythms with the institution that raised her. Sergeant Bobbie Draper travels to the world of her ancestors and Client Zero experiences a form of institutional psycho-violence conspired out of the Fanonian birth pains of a post-colonial state.

In the context of this text, these imprints partly facilitated each character's journey into a self-conscious epiphenomenal spacetime awareness. A latent field describes the relationships amongst changelings—in a spacetime with qualities of epiphenomenal time—to be inclusive of any temporal approaches. Epiphenomenal spacetime de-emphasizes the authority of causal experience of events, events are 'explained' when new information is illuminated by the shifting boundaries, or latent fields, connecting each changeling. Wright's linear progress narratives, which motivate modern nation state global organization, are approaches reflecting classical distinctions of space and time that still endure. Since the 20th century, Einstein's unification of space and time became the new classical model against which quantum theories continue to push. A major concern in contemporary physics is reconciling the inherent uncertainty of quantum reality with a deterministic but relativistic spacetime. Theories of quantum gravity and emergent spacetime are beginning to think about how to describe the underlying structure and origins of spacetime outside of a spaciotemporal framework.

UNITS OF MEASUREMENT
Jackie Karuti

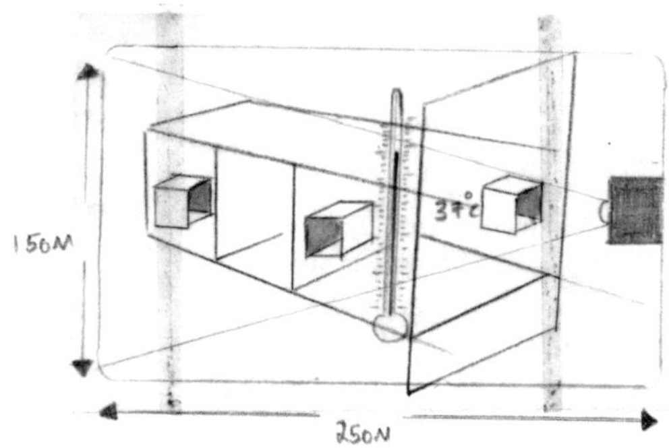

1. Length - meter (m)
2. Time - second (s)
3. Amount of substance - mole (mole)
4. Electric current - ampere (A)
5. Temperature - kelvin (K)
6. Luminous intensity - candela (cd)
7. Mass - kilogram (kg)
8. Work - joule (J) Work= Force x Distance
9. Volume - (cm3) Volume= LxWxH

A question goes: How does one consider all the possible times if you don't consider this time? I am reminded of deep time. A vast yet ambiguous quantum measurement which allows for other dimensions to open up while at the same time collapsing the classical and dominant understanding of time. Deep time rejects the modern understanding of time as being compartmentalized, measured and frozen in place. Deep time embraces the idea of duration as something that happens outside of us. A thing that unfolds slowly over extended periods of time causing gradual permanent changes that stretch out over a long period.

The living being has duration. It has duration precisely because it is continuously elaborating what is new and because there is no elaboration without searching. One can then say that durational practice is about how ideas come to us and that is over time, slowly & registered through unusual channels & frequencies that demand a certain awareness & intuition. Durational practice can be understood through slowness as defined by Tina Campt. That is, an overlooked velocity of living and by still-moving-images...as in, images that require the labour of feeling with or through them.

Like any rich and textured story, the idea of a durational practice complicates structure. It poses problems, does not claim to offer any answers, gives instructions, reveals a landscape, suggests tools for the journey, the weather occurs, strangers, animals and magical beings are encountered, measurements are taken, tasks are undertaken, physics is applied and perhaps we experience a revelation or revolution. As per Bergson, duration is the inner life of things. It can also be described as an indivisible act, qualitative effort, movement, continuous flow, progress, potentiality, essence, being and cinema.

Film vocabulary is layered with classification and taxonomic tools—exhaustive counting, cataloguing, lighting, lists, numbers—designed to organise what would otherwise be perceived as disorderly. Memorable characters are obsessed with witnessing, measuring and documenting animal decay. Animals on the whole are designed with a view of symmetry and one of decay's first characteristics is to spoil that symmetry.

Title	Hyènes (Hyenas)
Dir.	Djibril Diop Mambéty
Duration	1hr 50min
Year	1992

A Zed and Two Noughts
Peter Greenaway
1hr 55min
1985

I'm particularly struck by the overlap between two mysterious characters in Djibril Diop Mambéty's film *Hyenas* and Peter Greenaway's *A Zed and Two Noughts (Zoo).* In *Hyenas*, the once exiled Ramatou returns to her economically devastated village after many years away, only now richer than the World bank and thirsty for revenge. What's peculiar is that she dons gold plated prosthetics on her left hand and right leg claiming that she was the only survivor of a plane crash. In Greenaway's *Zoo*, the character Alba is the only survivor in a bizarre car accident involving a swan. She loses one leg and is fitted with an artificial limb. Later she has the other leg amputated as well. At one point she tells the story of a wealthy sex worker famous in Marseille who did not have both legs. In *Hyenas*, when Ramatou is banished from her village, the backstory of her whereabouts is unknown but she survived by making a living as a sex worker.

Both films explore love, acts of killing and the animal as metaphor but Mambéty goes where Greenaway hesitates; and that is to offer a critique of neo-colonialism and African consumerism through a palette of language and oral tradition relatable to the local context of Senegal where *Hyenas* is set. Greenaway's obsession with counting & numbers manifests itself in *Hyenas* through an economics of credit & debt that has compromised the town's ability to forge its future path. As for Mambéty, he lambasts the World Bank and the International Monetary Fund and portrays them as predatory financial institutions that continue to impose high interest rates in Africa's ongoing debt crisis. The association of the hyena with greed becomes a way of accounting for and finding form within these metrics.

At one point Ramatou remarks, "You don't wander the jungle without a ticket to the zoo". And when she's asked, *who can abolish time?* She promptly replies, "I can".

Taxonomy & naming: on looking, erasure and illegibility

What are you doing?
Making a zoo.
Can I look?
Only if you pay.

An act of attention implies solidarity between the mind and its object. There is a sudden realization that one hasn't learnt how to look after being asked to document what one sees. In calling the class to attention Moten says, "study is always happening long before the teacher enters the classroom & will continue long after he's left."

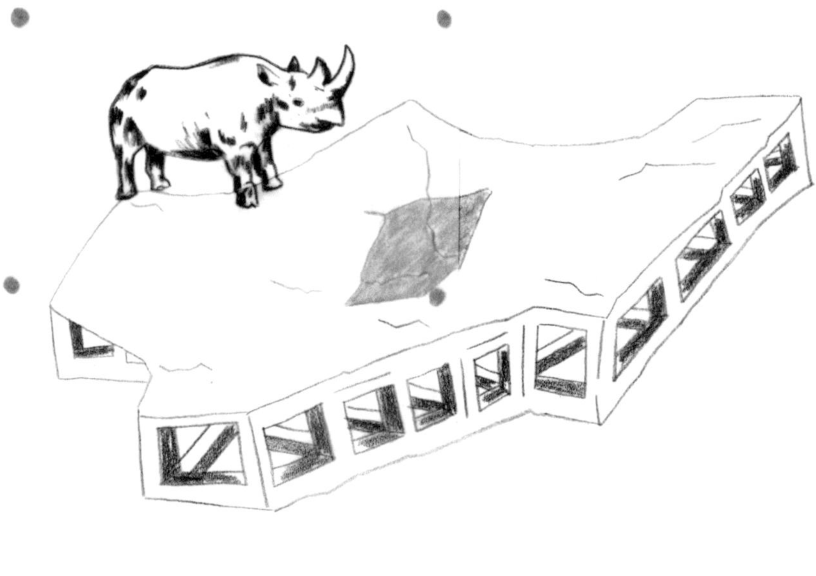

Measuring something that is alive is akin to eating something that is living or being. It's an interruption, cutting short and eventual end to life that limits, amputates and decimates generations in an all-consuming and planetary scale. It's a ruthless kind of grabbing & dispossession that blinds itself to the other species' right to existence & autonomy. It sees a thing as an object & not being sentient.

Dürer's Rhinoceros

Artist	Albrecht Dürer
Year	1515
Type	Woodcut
Dimensions	23.5 cm × 29.8 cm (9.3 in × 11.7 in)

Dürer's Rhinoceros is the name commonly given to a woodcut executed by German artist Albrecht Dürer in 1515. Dürer never saw the actual rhinoceros which was the first living example seen in Europe since Roman times. Instead, the image is based on an anonymous written description and brief sketch of an Indian rhinoceros brought to Lisbon in 1515. The king of Portugal, Manuel I had sent the animal as a gift for Pope Leo X but it died in a shipwreck off the coast of Italy. In the opera, *The Magic Flute* by William Kentridge and in the film *And the Ship Sails On* by Federico Fellini, one can see a direct reference to this particular moment perhaps inserted as a way to give weight and scale to the image they were both constructing.

Dürer's woodcut cemented the idea of how a rhino should look in people's mindsets but it's not an accurate representation. It depicts an animal with hard plates that cover its body like sheets of armour, a solid looking breastplate and what appear to be rivets along the seams. This can be interpreted as an image of war & conquest and speaks of the time. While advances in printmaking and the invention of photography eventually toppled Dürer's version, the rise of European travel to India and Africa and the emergence of public zoos allowed for more realistic illustrations.

Title	Taming of beasts. An extract from The Magic Flute	And the Ship Sails On
Dir.	William Kentridge	Federico Fellini
Duration	2min 31sec	2hr 8min
Year	2005	1983

Dürer's interpretation of the rhino remained the preeminent image in European culture up until the late-eighteenth century when a touring rhinoceros called Clara became a media sensation.

Miss Clara

Clara (c. 1738-1758) was a female Indian rhinoceros who became famous touring Europe for 17 years during the mid-18th century. She is considered the world's most travelled wild animal in captivity. In 1738, aged approximately one month, Clara was adopted by Jan Albert Sichterman in India after her mother was allegedly killed by Indian hunters somewhere in Assam. Sichterman was the director of the Dutch East India Company in Bengal at the time. Clara was later shipped to Europe in Rotterdam in 1741 becoming the fifth living rhinoceros to be seen in Europe in modern times since Dürer's Rhinoceros in 1515. She was known as the Dutch rhinoceros and received the name Miss Clara. She died in Lambeth, England on 14th April 1758 aged about 20.

Moo-o-o-o-o-o-o-o

"Monk entered the building, wearing what he wore in the shot (except for the rifle). The cow was standing in the vestibule... she had not yet taken her place on the set. Monk went over and put his arm around her shoulders. He bent down close to her right ear and very calmly and quietly said: 'Moo-o-o-o-o-o-o-o.' The cow seemed unimpressed. Monk then just walked onto the set; sat down at that battered upright piano and proceeded to play for about an hour and a half. The piano was terribly out of tune and I'm sure didn't have all of its keys. But it didn't matter...it was great! After the photo session Monk got up and left with The Baroness. The only word he had spoken in all that time was to the cow."

—music producer Michael Cuscuna, recalling a story told to him by Columbia Records' former art director Richard Mantel on the story behind Thelonious Monk's *Underground* (1968) album cover.

Grévy's Zebra

The Grévy's zebra was first described by French naturalist Émile Oustalet in 1882. He named it after Jules Grévy, then president of France, who in the 1880s was gifted a zebra by the government of Abyssinia. Their stripes are narrow and close-set, being broader on the neck and they extend to the hooves. The belly and the area around the base of the tail lack stripes and are just white in colour, which is unique to the Grévy's zebra. It is found in parts of Kenya and Ethiopia.

Scientific name	Equus Grevyi
Genus	Equus
Species	E. grevyi
Weight	350-450kg
Size	125-150cm
Gestation	13 months
Life span	12-13 years
Habitat	Grassland and savanna
Diet	Herbivorous
Predators	Lions, cheetahs, hyenas, leopards, African wild dogs, humans
Conservation status	Endangered (Population stable)

Besides the Grévy's zebra, the plains zebra is the most common and geographically widespread species of zebra. Six or seven subspecies have been recognised, including the extinct quagga which was thought to be a separate species. The plains zebra subspecies include:

Maneless zebra	(E. q. borensis)
Grant's zebra	(E. q. boehmi)
Crawshay's zebra	(E. q. chapmani)
Chapman's zebra	(E. q. chapmani)
Burchell's zebra	(E. q. burchellii)
Selous' zebra	(E. q. selousi)
Quagga	(E. q. quagga)

The quagga is an extinct subspecies of the plains zebra that was endemic to South Africa until it was hunted to extinction in the late 19th century. Charles Darwin deemed the quagga a separate species but early genetic studies have supported it being a subspecies of the plains zebra. (Equus quagga). The quagga was extinct in the wild by 1878 with the last remaining mare residing at the Amsterdam Zoo between 1867-1883. It died there in August 1883 and with her an entire species. It was not realised until later that she was the very last of her kind.

The animal has memory but no memories
— Giorgio Agamben

The quagga resembles a chimera of a horse and zebra. The name *Quagga* was taken from a native language in Southern Africa. It is supposed to be onomatopoeic as the sound made by pronouncing the word reflects the call made by the animal. Quagga exhibits that are currently held in museum collections are not fossils but actual skeletons as these animals have only become extinct in recent history. The Quagga Project is an attempt by a group in South Africa to use selective breeding to achieve a lineage of Burchell's zebra which visually resembles the extinct quagga.

Details about the last remaining Quagga residing at the Amsterdam Zoo

Scientific name	Equus quagga quagga
Sex	Female
Head-body	2.28 m
Tail	0.46 m
Ear	0.6 m
Hindfoot	0.46 m
Shoulder height	1.20 m
Residence period at the Amsterdam Zoo	9th May 1867 - 12th August 1883
State of preservation	Very good

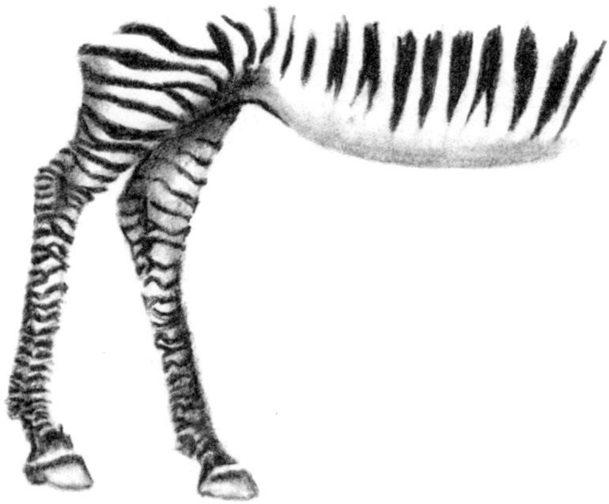

A drawing of a Grévy's zebra being looked at. Location: ARTIS, Amsterdam Zoo, 2024

Q: Why does the classification of species disregard indigenous ways of naming & knowing?
Q: Is a zebra white with black stripes or black with white stripes?
Q: Is the zebra emerging or disappearing?
Q: Is the drawing (not yet) complete?
Q: Are we drawing or erasing the zebra?
A: But isn't erasure a form of drawing?

Encounter with a wounded buffalo–an excerpt from a letter written by Rosa Luxemburg in prison to her friend Sophie:

"... In the courtyard where I walk, army lorries often arrive, laden with haversacks or old tunics and shirts from the front; sometimes they are stained with blood...The other day one of these lorries was drawn by a team of buffaloes instead of horses...so overladen indeed that the buffaloes were unable to drag it across the threshold of the gate...At length the buffaloes succeeded in drawing the load over the obstacle, but one of them was bleeding. You know their hide is proverbial for its thickness and toughness, but it had been torn. While the lorry was being unloaded, the beasts, which were utterly exhausted, stood perfectly still...I stood in front of the team; the beast looked at me, the tears welled from my own eyes... The driver, hands in pockets, was striding up and down the courtyard, smiling to himself as he whistled a popular air. I had a vision of all the splendour of war!..."

sb-foot, sb-step, sb-ell

The atomic world is a ghostly abstract place that crystallizes into sharp existence only when we set up a measuring device to interact with it. Unlike a classical object that always occupies a definite position in space, an electron could be in multiple places at once until the moment it is measured. The natural world at a microscopic scale means that something can be spread out in space and the best we can do is work out a set of numbers that give the probability of finding it, not at a single location, but at every point in space simultaneously. Depending on the point where it was last seen we can calculate and estimate different positions and times that allow us to assign probabilities to where it might show up next. It is however impossible to set up a single experiment in which we can measure, as accurately as we wish, both where an electron is and how fast it is moving, simultaneously.

My first encounter with stanley brouwn's work was years ago through a text that described him as an elusive artist whose practice was distinctly marked by his own absence. I was confused by this definition. The insistence on being invisible or the refusal to be made legible made searching or trying to decipher his work almost impossible because he objected to his work being documented or reproduced in print.

I was therefore elated when I recently chanced upon his work for the first time inside the Stedelijk museum in Amsterdam where a modest collection of his work was on view. I will attempt not to describe the work because anyway, it was nearly impossible to take in its rich sparseness in the excess that is the contemporary art museum today.

What immediately struck me was how sanitary and minimalist the wide single room looked with its subdued shades of white and grey and the cautionary, *Photography not allowed* sign at the door. There was an ensemble of drawings, books, pamphlets, sheets of paper marked with measurements, directions, distances, numbers and objects methodically displayed on glass table-tops. In some cases, what seemed like blank sheets of framed off-white paper was actually marked with letters and numbers once you moved closer. On one wall, a table was piled with grey filing cabinets. A few were open and stacked with index cards or blank sheets of A5 size white paper rendering the room in a very administrative light. The spacing of everything and how one paced across the room also seemed intentional. One of the titles indicated that; *"at this moment the distance between stanley brouwn and yourself is x feet,"*. It didn't point to any particular artwork.

In lieu of standardized measurements such as the metre, brouwn developed his own personal measuring system based on his own body such as his hand, arm and foot and also incorporated old units of measurement such as the foot, the step an the ell—sb-foot, sb-step, sb-ell. He recorded his own footsteps in various cities on index cards that he then stored in grey metal filing cabinets. His measurements of the places & routes he travelled framed walking as a way to exhaustively record and map out the lengths of the world and objects around him.

In some cases, he imagined the distance between himself or more precisely his absence in relation to space producing works titled as; *"at this moment the distance between c and yourself is x feet,"*. This unique form of abstraction not only directs how people move and look but also reduces the rational lines in his work to traces and measurements of a body, while viewing things in relation to one another rather than according to absolute models.

When brouwn moved to Amsterdam from Suriname, he encountered the Zero movement; a group of artists who sought anonymity and rejected the evident authorial signature. This early influence helps to explain why he would later go on to explore the dematerialisation of art and the disappearance of the author. *This way brouwn* consists of the artist stopping passersby in the street and asking for directions, which he has them scrawl on a piece of paper. The results were experiments in cartography that featured markings and street plans sketched to accompany verbal directions. The artist then stamped the phrase THIS WAY BROUWN on the documents to authenticate them as artworks. This gesture allowed brouwn to place the act of creation into the hands of others therefore complicating what authorship means in art making.

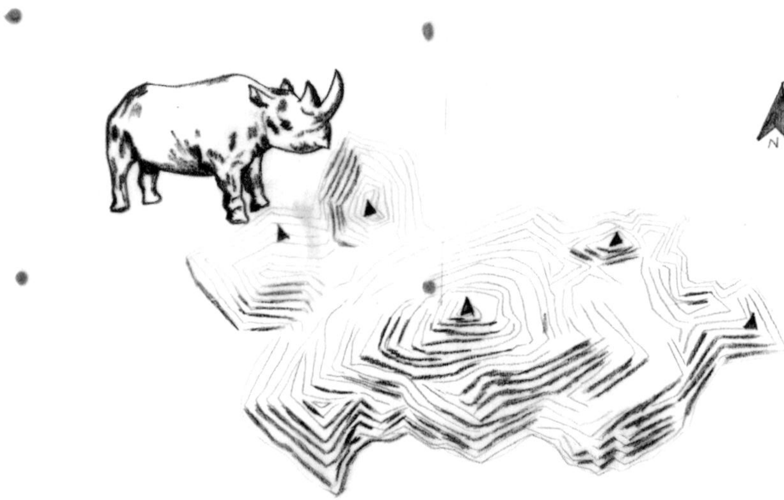
N

The reading group: from glossary to garden

– Like electricity; to generate is to create or produce what mining extracts. It also means to regard something collectively. i.e, generation. With minimal tending and interference, the reading group is to become a living thing or environment that (re)generates and mutates to spawn chaotic & unthinkable things that embody what it means to be destructive, invasive, undesirable and monstrous. Flowers. Weeds. Bugs.

– A reading group by nature collates what is read, written and mediated upon and assumes a logic to what is presented. However, by inviting the disruption and fragmentation of study, the reading group by way of the glossary allows for uncertainty through what is generated while further obscuring our understanding of authorship.

– So, what is the life cycle of weeds? Annual weeds germinate from seeds in the spring, to fully bloom in the summer, before dying again in the winter. Tulips for example need 8 to 16 weeks of artificial winter. After bringing the plants into temperatures resembling spring, the tulip will sprout along with leaves quickly emerging. A flowering plant will then appear within 15 to 30 days.

– Rather than explaining the meaning of a word, we seek to follow Tina Campt's definition of a glossary; i.e *"...a series of keywords crucial to thinking with, through and alongside contemporary articulations of black visuality."*

– Meeting times: 1hr every Wednesday, 7pm. No sign up needed.

- Come with your 'term /keyword' & present it. It will be added to the glossary:

Wander lines	Dignity	Night school
Respair	Navel string	Translation
Foam	Refusal	Cut-
		~~Bounderies~~

– Anyone in the group is welcome to edit.

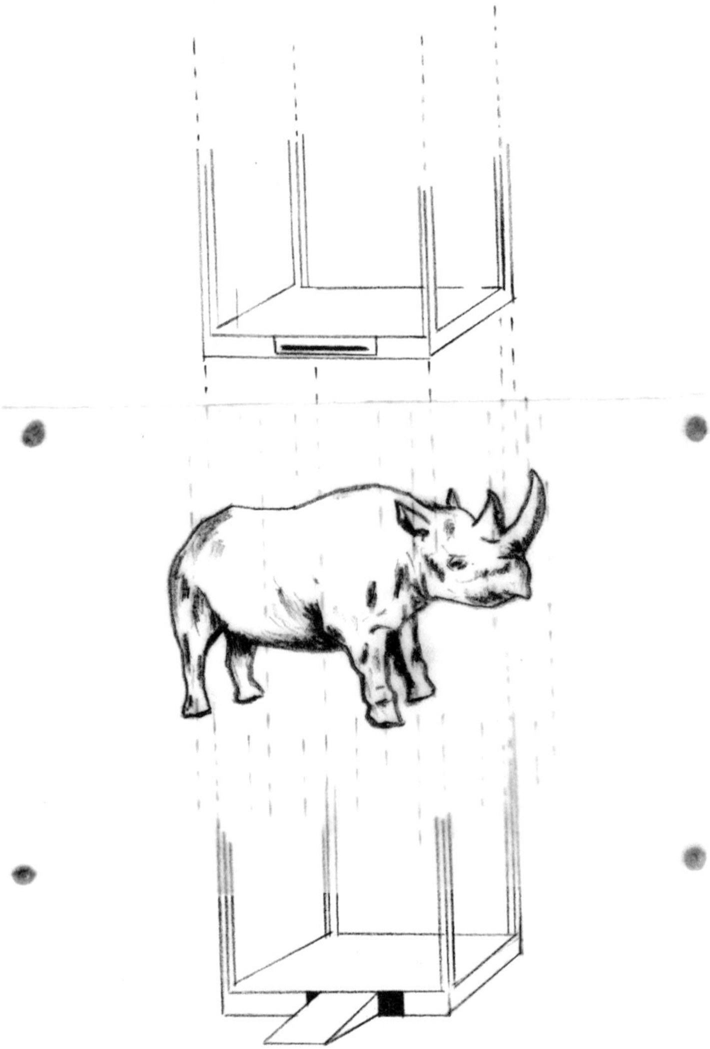

250×150cm: intercepting images in space & time

In order to look, one must construct an instrument or a machine. One that cannot be defined so much as a thing but what it allows. And this is the ability to act as a site of constructive repair. Through the hands of the Engineer and the eyes of the Chemist, the function and usefulness of both the projector and microscope are tested in two video experiments. The Engineer identifies the problem with a faulty projector to be a weak lens. He tests the strength of the beam as he explains the process in different languages. The chemist looks through a microscope into a universe of cell division and proceeds to test a sample. A moment of alchemy occurs when both witness the forming of a cloud through a haze of movement and blurriness in these acts of testing and looking. This offering of a world is similar to what cinema does. It situates the projection & magnification of the image as a telescopic leap, giving us a macro view of ourselves as a species while allowing us to see how one is alienated by what they encounter.

Dir.	Jackie Karuti
Title	The Engineer
Duration	4min 16sec
Year	2019

Jackie Karuti
The Chemist
3min 47sec
2019

I used to love watching films using a projector in my old tiny studio but kept encountering a problem. Due to the small space, the throw distance was not ideal and sometimes did not allow for a great viewing experience. The projector would also heat up the room so naturally, I longed for a more adequate machine or to move to a bigger space. This would allow ample space between the projector and the image or more precisely it would increase the distance between myself and the image. In my quest to solve the problem, I obsessively started measuring and mapping this specific area arriving at a viewing acreage of 250×150cm. However, the more I made these map drawings, the more I realized that I would erroneously put down the dimensions as meters instead of centimeters. So sometimes 250×150cm would appear as 250×150m is some drawings.

As if this was not already confusing to me and those who acquired some of these drawings, I complicated the problem even further by acquiring a two-acre piece of land approximately 145km from my studio. In my occasional out of town getaways, I had discovered this land as well as a vast seasonal river a few meters from the site. It took me numerous visits over the following years before I came to the conclusion that my reason for acquiring the land was simply to situate it as an observatory.

Studio dimensions: 250×150cm
Nairobi to Ngatataek: 145km
Land dimensions: 2 acres (approx. 87120 square feet)

The duration needed to transform 250×150cm into 87120 square feet takes 145km or 32 hours of walking.

When I finally got back to the studio, I realized that I had solved the problem. By encountering and acquiring this new space miles away, I had projected the image further-out; collapsing space & distance. In my desire to increase the size of the space or achieve a sharper and more legible image, all I had needed to do was exit the studio and encounter the world similar to what cinema allows us. The minimal confines of the studio could no longer accommodate the image I was seeking or the space I had been trying to construct. It existed in the erroneous cartography and immeasurable ambiguity similar to the shapeshifting seasonal river I had encountered.

A THEORY APPARATUS FOR OBSERVATION

Rohini Devasher & Sabih Ahmed

Astronomy is unique by virtue of being the oldest of the observational sciences. It is also one of the few scientific fields where there will always be a need for the amateur because the sky is immense and there are simply not enough people looking at the stars.

The practice of observation unfolds in time while also 'folding-in' time. The tools of observation focus attention and assist in both the measurement and the making of time. They set provisional boundaries to what would otherwise be an overwhelming field of view.

In the days before digital applications such as Stellarium, Celestia, Sky Map and Space Engine, which allow the user to point their phones at the sky and identify any object visible and invisible to the naked eye, observers would use an analog star map, a planisphere. A planisphere consists of two rotating disks, one a star chart, the other an opaque circular overlay with a clear window mounted on a common pivot. By adjusting these disks, the star chart could be aligned to show the visible stars for any given date and time. For the amateur astronomer, it is an apparatus to help in learning to recognize stars and constellations; how to 'star jump'; how to navigate the night sky. An object of astonishing simplicity grappling with enormous scales of space, time and measurement—it is an instrument of time travel.

This text is inspired by the workings of a planisphere, where reflections pivot around observations gathered by the authors since 2009. The text comprises vast arcs of thought starting from various observer's statements and setting off into reflections on light, darkness, gravity, perception, images, and a bio-technic understanding of observation as a distributed apparatus.

The Speed of Darkness

OBSERVER:
"The eclipse is a very powerful symbol. In many ways.
It's where black dominates.
Light would have no meaning if there were no dark.
Because it is not the light, the Sun you see every day, far brighter,
and that doesn't do anything to you.
It's the black in the middle that does it all.
People say that the black does it because it's blocking out the light
and you suddenly realise how important the light is.
But I said, no! No eclipse blocks light.
Night has more dark than an eclipse.
It's just that the black suddenly says 'I have a voice'."
Raj Shekhar, interview during the 'City as a Studio' fellowship, New Delhi, 2010

REFLECTION (Sabih): The phrase 'the long shadow of history' suggests that there is a source of light casting shadows. How is it that history becomes that source of light against which every other story is silhouetted? And who imbued history with such light?

Georges-Didi Huberman's 'Survival of the Fireflies' put it succinctly. The glorious light from Dante's 'Paradiso' all the way to European Enlightenment is not a metaphor for knowledge but of totalitarian power. It is the dictator that blinds with fierce lights while the people glimmer in darkness like fireflies. Much of the world is made up of illegitimate inheritors of the Enlightenment, condemned to prove that they understand the inheritance and can hold themselves up to it. They must keep producing themselves before this ray of light.

However, what does a paradigm of darkness tell us? What knowledge and techniques of observation emerge from conditions of darkness and disappearance—when something does not appear, rather, refuses to appear? Conditions of appearance have always been premised on continuity—light moves as a continuous beam separating the visible from the invisible and truth from oblivion. These are continuous lines that best not break. Be it the converging lines in a

Renaissance painting, the light passing through a camera aperture, or soft spotlights in museum vitrines that rescue artefacts from the shadows of antiquity, or, the surveillance regimes enforcing upon populations an uninterrupted visibility.

A theory of perception premised on conditions of darkness calls upon technologies, techniques, defiances, and parameters of encounter against the regimes mentioned above. It calls for disordering the ontology of darkness, where disappearance is not the opposite of appearance, where shadows are not the absence of light, where representation is not the cure for a lack.

A paradigm of darkness does not call for rescuing, excavating, or trying to see that which has vanished. It is to reorder the very paradigm of perception and all its incumbent techniques, namely seeing, observing, examining, scanning, graphing, surveying, surveilling, staring, gazing, projecting and so on.

To start with the condition of darkness is to start from an abyss. An abyss that is historical and constituted by Enlightenment. Édouard Glissant refers to it in the opening of 'Poetics of Relation'. An abyss that partakes of over two centuries and over thirty million people enslaved, deported, and subjected to an unknown. For Glissant, the abyss is three-fold. The first is in the belly of the ship, the second in the depths of the seas, and the third a reverse image of all that was left behind and cannot be recovered. It is an abyss at the bottom of the Atlantic and countless other oceans. Glissant would write that the "experience of the abyss lies inside and outside the abyss".[1] Many philosophers have extolled the limits of what is knowable, the lacuna in testimony. But Glissant says "the absolute unknown, projected by the abyss and bearing into eternity the womb abyss and the infinite abyss, in the end became knowledge... People who have been to the abyss do not brag of being chosen. They do not believe they are giving birth to any modern force. They live Relation and clear the way for it..."[2]

To think of the abyss is to consider what it means to observe darkness, i.e. to observe the dark and to observe in darkness. It is an occasion to raise questions about the ethical dimension of observation itself. What does it mean to observe, and what are the techniques, technologies and assumptions built into our understanding of

1 Glissant, Édouard, *Poetics of Relation*. Translated by Betsy Wing, University of Michigan Press, 1997.
2 Glissant, Édouard, ibid.

observation? And what does it take to observe that which is unknown when the foundations of knowing are built on a history of obliterations. It requires a reform of perception, and therein a reform of epistemology and the practice of observation itself.

Putting aside the age-old question of how fast light travels, the question that looms large after seeing an eclipse is, 'how fast does darkness travel?' What theory of perception emerges from conditions of darkness?

REFLECTION (Rohini): July 22, 2009, the longest solar eclipse of the century, and the Sun refused to appear, or rather was completely occluded from sight. Perhaps because of the complete dark, it was a violently vertiginous feeling, as though I was both on and off the planet simultaneously. The people around me were similarly caught between silence and screaming, between shock and wonder. But that wonder was not the wonder of the sublime or of saccharine loveliness, but of the dark, of the deep, that wonder had teeth.

In her book 'Reclaiming Wonder: after the Sublime', Genevieve Lloyd says "Wonder, in its many forms, depends but also thrives on the absence of certainty. It has often been associated with explicitly celebrated forms of 'not-knowing'".[3] Historian of science Lorraine Daston in her article 'Wonder and the Ends of Inquiry' describes wonder, horror and terror as passions of the unnatural.

"Horror, terror and wonder are triggered when a major disruption of order (whether moral or natural or both) is registered as such: an act of perception and judgement that presumes some acquaintance with the particular sort of orderliness that has been breached—that, as the phrase goes, 'something is not right.' Subterranean connections bind the passions of the unnatural to each other."[4]

If wonder thrives on uncertainty, then the speculative or horror thrives on the improbable. That moment when the unexpected comes in and forces you to pay attention. Turning something on its head, asking the question 'what if!?' A question that is at the heart of so much science fiction and speculative fiction.

Then there is the strangeness of light as demonstrated by the famous double-slit experiment which many say is at the heart of the deep strangeness of quantum mechanics. Very briefly, the double slit

3 Lloyd, Genevieve, *Reclaiming Wonder: after the Sublime*, Edinburgh University Press, 2018
4 Lorraine Daston, *Wonder and the Ends of Inquiry*, Examined Life Magazine, 2014

experiment was first used by British polymath Thomas Young in 1801 to demonstrate the wave-like behaviour of visible light. In modern physics, the double-slit experiment shows that light and matter can sometimes behave as waves and sometimes as particles. This inconsistency is considered as evidence for the fundamentally probabilistic nature of quantum mechanics.[5]

This used to be known as wave-particle duality, but is now more commonly referred to as the 'measurement problem' in Quantum Mechanics. It is a problem because, as David Wallace—author of 'Philosophy of Physics'—says, the theory requires inconsistency... "We seem to have to shift, inconsistently, from one description to another and back, depending on what features of the experiment we wish to describe."[6] In quantum mechanics, the measurement problem is the problem of how / whether, when 'observed', wave function collapse occurs. In simpler terms it is when all the probabilities collapse into a single state due to interaction with the observer / world.

There is a wonderful parallel for this entangled relationship between observer, world, object problem / inconsistency in the work of author Jeff Vandermeer, specifically his 'Southern Reach' trilogy (2014), where the 'natural' world is both under threat and threatening. The plot begins as many other works in the sci-fi genre do with a seemingly first-contact encounter. Except this encounter is with a zone, an entire landscape, called Area X, the boundaries of which are marked not by a border but a 'shimmer' which keeps expanding. This shimmer has, we later discover, a horrifyingly refractive and prismatic effect on everything including DNA within its reach. Fungi and plants fruiting people shapes, a bear who screams in the voice of its human victim, a dolphin with almost human eyes...

The Field of Vision

OBSERVER:
"A common misunderstanding is that when you observe visually the experience is limited to yourself, and when you pursue astrophotography, you can share the result with the world. Astrophotography shows you what you cannot see, colours of nebulae, faint details on planets. But not many people have experienced an observation with a large

5 https://en.wikipedia.org/wiki/Double-slit_experiment
6 Wallace, David, *Philosophy of Physics, a Very Short Introduction*, Oxford University Press, 2021

telescope on a mountain peak. Take a 20-inch telescope to Hatu peak, and observe visually, you will forget astrophotography, and remember the sight through the eyepiece forever."
Ajay Talwar, The Night Sky Observation, Alwar Rajasthan 2019

REFLECTION (Rohini): In the hours between 6 pm October 5 and 6 am October 6, 2019, I curated an overnight observation in collaboration with astro-photographer and fellow amateur astronomer Ajay Talwar. The site of the 'Saros 132 - Night Sky Observation'[7] was an observatory Ajay had designed and set up at the Sagar School Observatory in Alwar, Rajasthan. The idea was to bring together a group of people, some with experience of astronomy and some without, to think together about what it might mean to observe observation in practice. Ajay planned an astral itinerary that included the Moon in its first quarter, 7 days old, 48% lit in the constellation Sagittarius; the planet Jupiter, its four Galilean satellites, the storms that formed cloud bands on its surface; Saturn; Uranus; Neptune; and a navigation through star clusters and a dance of the double stars within the Albireo system 433.8 light years away. Of all the entities we observed that night and early morning, it was Saturn that took centre stage.

The distance to Saturn from planet earth is constantly changing. When the two are closest, they lie approximately 1.2 billion kilometres apart. At their most distant, when they lie on opposite sides of the sun from one another, they are over 1.7 billion kilometres apart. When we look at Saturn, we are looking about 79 minutes into the past.

There is a strange sort of distance between the virtual and the real. Hubble, Cassini, James Webb and their marvellous images have found their way into everyone's lives and taken up places in our imaginations. Entities have been assigned forms, shapes, and thereby, meanings. But the experience in the field with a telescope or binoculars is very different. I once asked an amateur astronomer how he deals with the anxieties and false expectations that often accompany a first-time observation. That moment of 'is that all?' His answer was, you have to remind them of where they are, and of

7 *'Saros 132: Night Sky Observation,'* an incident by Rohini Devasher in collaboration with Ajay Talwar, as part of 'Five Million Incidents' organised by Goethe-Institut, Max Mueller Bhavan, New Delhi and Kolkata, conceived by Raqs Media Collective, 2019-2020

what they are seeing - that the light of that star has travelled so far before pinching the light receptive photons in their eye. As an experience, it is visceral, it is 'real' and it is uniquely theirs.

REFLECTION (Sabih): The practice of observation reminds us that seeing is never unmediated. The technologies and techniques through which we see produce a fantasy of unmediated seeing, as if there is a form of experience that is somehow more real, instinctive or natural if those technologies were put aside. The question of whether or not our experience of the world can be unmediated is fundamental to larger questions of what it means to be human. It raises questions about whether human experience can ever be purely biological when everything we know about the world and the universe is only through techniques of observation. How else would we even know what we are seeing?

Arab navigators were known to use a single or double finger width, their thumbs and little fingers on an outstretched arm, or an arrow held at arm's length to sight the horizon at the lower end with Polaris at the upper.[8] An observation was made using the geometry between lines of sight, the angles of digits and the alignment of the earth with stars. The technique depended on the angle between the horizon and the Pole star to give information about the altitude of one's position.

Nicholas Mirzoeff in 'How to See the World' writes, "[i]nstruments like the astrolabe, quadrant, and sextant refined this way of gaining orientation by using the horizon and the stars. There have been several periods of dramatic change in the visible world before ours. The nineteenth century was famously described by historian Jean-Louis Comolli as a 'frenzy of the visible' because of the invention of photography, film, X-Ray, and many other now forgotten visual technologies in the period."[9]

We know that the development of maps, microscopes, telescopes and other devices made the 17th century another era of visual discovery. In more recent times, the older forms of visual codifications have been surpassed by newer forms. Wendy Hui Kyong Chun has written elaborately on computation and image-making, "when the computer does let us 'see' what we cannot normally see, or even

8 Ifland, Peter, *The History of the Sextant*, October 2000 (https://www.mat.uc.pt/~helios/Mestre/Novemb00/H61iflan.htm)
9 Mirzoeff, Nicholas, *How to See the World*, Basic Books, 2016

when it acts like a transparent medium through video chat, it does not simply relay what is on the other side: it computes".[10] An ultrasound scanner measures the inside of a person's body using sound waves digitally rendered to form an image. In most cases, we see in the image what we could never see with our own eyes. It continues the long legacy of our ways to see the world enabled by broader techniques of sensing.

The Cosmos as a Compound Eye

OBSERVER:
"*Gravitational lensing can also function like a huge magnifying glass, allowing us to see galaxies that otherwise may not be observable. These galaxies are also further back in time allowing a glimpse into the distant past. Gravitational lensing allows observation of large cosmological distances.*"
Prof. Ajith Parameswaran, interview at ICTS – TIFR, Bengaluru, 2023

REFLECTION (Rohini): What happens when forces themselves function as techniques of seeing? If we think about the Einstein Ring for instance. Strong gravitational fields such as those created by black holes or galaxies can function as giant lenses that produce a warping in the fabric of space-time, resulting in extraordinary effects such as multiplication, magnification, distortions, etc. If the source, gravitational lens, and observer are all in perfect alignment, the light from a distant galaxy may appear as a ring. A perfect ring in space.

REFLECTION (Sabih): To regard cosmic forces as techniques of seeing is to extend one's sensing capacities not only into technical prosthesis like spectacles or telescopes but also into larger phenomena. To practice it, one cannot assume standing at a distance for observing, but rather conjoining oneself with the environment in order to sense. To form agile integrations between biological and ecological capacities. This lies at the heart of being immersed in the world.

In art history, J.M.W Turner's figuration of the landscape and the sea involved that kind of immersion as opposed to distance.

10 Chun, Wendy Hui Kyong, *Programmed Visions: Software and Memory,* MIT Press, 2011

His intent was to immerse the spectator. This immersion is at once a literal place of viewing from amongst the waves, being captivated by the power of the scene, and a metaphorical deployment of a space of transition between regimes of power. In short, being immersed was to be in crisis.

Now regard this in view of the cosmos. What is the jurisdiction of a cosmic event? Does it stop at the earth's atmosphere? Or does the earth share the same proscenium wherein biological actors are in interaction with cosmic actors to tell a story.

Modern science and a colonial deployment of technology has celebrated science's ability to conquer nature, conquer distance, space and time. Practices of observation exceed the colonial institutions in which they are often located. They invite you into discourses of proximity, entanglements and simultaneity that are not a conquest but a dance. James Baldwin famously wrote in 'No Name in the Street', "[t]here is a reason, after all, that some people wish to colonize the moon, and others dance before it as before an ancient friend."

REFLECTION (Rohini): In 2016, I was invited to be part of a show at the Spencer museum of Art in Lawrence, Kansas, and while I was in Lawrence I interviewed a group of Storm Chasers. I was interested in the possible connections between the storm chasing community and the amateur astronomy community, one so enthralled by the night sky, the other by extreme weather, could tell us about the nature and complexities of meaning-making? Amateur astronomy is as much about an encounter with the self as it is an experience of space and horizon. Similarly, weather because of its scale provides an almost mythic realisation of oneself within an environment.

The obvious link between amateur astronomers and storm chasers would seem to be a shared experience of the infinite and the transcendent. Reclaiming a lost sense of awe, a sense that they are part of something greater. What perhaps separates them is the experience of terror. Almost without exception every storm chaser I interviewed had had a deeply traumatic experience with a tornado as a child which resulted either in a burning desire to try and understand the phenomena as a means to ensure future prediction and control or an equally driving need to come as close, physically close, as possible to a childhood horror.

What is also distinct is their understanding and perception of time and of transformation. Astronomers draw comfort from the unchangeable nature of what they observe. Storm chasers on the other hand both expect but also dread the unpredictability, uncertainty and mutability that is central to extreme weather. Time functions differently and is experienced differently.

REFLECTION (Sabih): The field of vision is always reconstituted by techniques and technologies of its time. And yet, both are also simultaneously out of joint with their own time. They may produce the experience of time-space continuums, but also discontinuums. A practice of observation is therefore one that always has to take on the task of defining what a field of vision is, i.e, the conjunction of technologies, techniques, bodies and sites mobilised. The constellations we observe out there are constituted by constellations over here. The field, as one might otherwise assume, is never already present. It is produced by the conjunctions that also produce the conditions of observation. A critical practice of observation would therefore be a critical undoing of the relation between the technologies, techniques, the site and the observer, none of them being always-already given. A practice of observation directed towards the cosmos opens an interesting and curious situation where the figure of the practitioner is a subject that is situated within the object of their inquiry. We are within the cosmos that we are trying to observe, not removed from it. This radically changes an older and dominant paradigm of scientific observation where the subject and objects were kept at a distance.

OBSERVER:
"What if you tried to apply quantum mechanics to the entire universe? Then there are no external observers, and in a theory of gravity it's true that you cannot really separate a system from things outside a system. Then you could ask, what do measurements mean when you are yourself part of the system?"
Prof. Suvrat Raju, interview at ICTS-TIFR Bengaluru, 2023

REFLECTION (Rohini): In quantum mechanics, if it is true that the nature of observation changes with the acts of the observer, what does this mean for changing methodologies of observation and the role of the observer? Are observations measurement or interactions? Is this question even relevant? String theorists hold that observations are always interactions. Some go as far as to say that what is an observation in one perspective is an introspection in another.

A field is many things. When understood within the frame of sight, the field of view is the extent of the observable world seen at any given moment. It is the area of inspection captured on the camera's imager. In astronomy, the field of view is usually expressed as an angular area viewed by the instrument. In quantum mechanics, fields are mysterious, fluid-like substances that sometimes act like particles and sometimes like waves. And so, depending on the observer, the field is sight, resolution, measurement and interaction all at once.

What is the amateur astronomer's field? Is it the observatory? Is it the site where they set up to view an eclipse or meteor shower? Omar Nasim in his essay on the observatory describes the threads that bind the laboratory, the observatory and the field.

"The laboratory is characterized by its sterile and standardized placelessness, one that is conducive to universally replicable findings. The field, on the other hand, is often distinguished for its ruggedness and site-specificity, where local variations are themselves made the subject of study. Without being reduced to a border zone,[11] the observatory has aptly been described as a middle ground between laboratory and field,[12] and as a unique site where characteristics of both field and lab co-mingle.[13] Aside from the deconstruction of universal placelessness into multiple intersecting places, sites and localities, let us not forget that just as vital are the strategies and techniques employed by scientists for transcending, erasing, or silencing locale specificities; or what Peter Galison has called the local de-localization of scientific work.[14] The observatory as a hybrid site provides historians of science with an exceptional opportunity to examine more fully the multiple dimensions of both directions: from the placeless to the situated, and from the local to the de-placed. After all, those who worked at observatories directly faced the ideals and challenges of both the laboratory and the field. Indeed, permanent observatories were de-placed, as they became nodes in socio-political, mathematical, and geographic networks."[15]

11 Kohler, Robert E., *Landscapes and Labscapes: Exploring the Lab-Field Border in Biology*, University of Chicago Press, 2002

12 Le Gars and Aubin, *The Elusive Placelessness of the Mont-Blanc Observatory (1893–1909): The Social Underpinnings of High-Altitude Observation*, 2009

13 Lane K. Maria D., *Geographies of Mars: Seeing and Knowing the Red Planet*, University of Chicago Press, 2010

14 Galison, Peter, *Image and Logic, A Material Culture of Microphysics*, University of Chicago Press, 1997

15 Nasim, Omar, *Observatory*. In: Sommer M., Müller-Wille S., Reinhardt C. (eds) Handbook of History of Science. JB Metzler, Stuttgart, 2017

The site of our collective observation in Saros 132 was on the campus of a boarding school in Alwar, Rajasthan, about 3 hours away from Delhi. The observatory designed and run by Ajay Talwar would host weekly classes for students of the school using the research level 14-inch Celestron Schmidt Cassegrain telescope. The observatory serves several categories at once—as an observatory, a laboratory for students, a field, and for those who made the visit that evening, a field station. Saros 132, as an incident, invited a collective observation on observation itself. The observatory became the site, the medium, and the incident.

What is the Resolution of the Universe?

OBSERVER:
"Saturn is one billion kilometres away, and you see Saturn because the light reflected off Saturn from the Sun reaches your eye, and it is quite a moment. We could see the shadow of Saturn on its own rings because of the way the Sun was positioned vis-a-vis Saturn. But that object called Saturn, or what we thought Saturn was, as known through models, photographs, etc., is not the Saturn we saw. What we saw, that quiet little object in the sky, the time horizon of it, to even think about 1 billion kilometres is a bit complicated. You have to unlearn rapidly what you have seen before to even make it intelligible. So, what happens is there is a search for language, you have to unpack your unseeing of the world, and you have to make intelligible what you saw. Which as an experience was very frugal."
Jeebesh Bagchi, The Night Sky Observation, Alwar Rajasthan 2019

REFLECTION (Rohini): Observation is closely linked to the idea of the collective. The collective as community but also collective perception. One of the things that stands out for me about our observation was how the night was divided into sections of time, narrowly focused (literally) around an object. The observation schedule was carefully planned by Ajay and myself, and we orbited around it both inside and outside the observatory.

The study of early drawings of nebulae in the 19th century have shown that each representation was in a sense a kind of composite. Omar Nasim describes them as a repository of data derived over a period of time. He writes, "(i)t was also a slowing down of the accumulation of a detailed final pictorial representation—an image that was a collection, a composite of information derived on many different nights and days of work. The slowing down in the procedures of observation, thus, resulted in an extended, protracted, and continuous gaze on a faint, barely visible object."[16]

For those of us observing that night, I think time was similarly stretched and while we did not draw (at least not on site), something similar happened with memory. No two people saw any object the same way. In fact, we kept comparing what we saw and drawing correspondences.

REFLECTION (Sabih): In 2012, NASA released an updated version of the Blue Marble photograph taken from a distance of merely 825 kilometres from the Earth's surface. This stands in stark contrast to its 1972 counterpart that was taken from approximately 29,000 kilometres away as a single snapshot. The new image on the other hand was obtained over a period of eight hours and six orbits of the earth. The staggering difference between the time and distance for taking these pictures is telling.

The new Blue Marble is a composite image from a large set of digital images produced by the satellite Suomi NPP. From the satellite's orbit, a full view of the planet is not possible. One has to go approximately 11.5 kilometres further away before the entire globe can be seen. The 2012 image is carefully constructed in order to give the illusion that it was taken from one location at one moment in time, whereas it was far from the case. Mirzoeff describes this as "a good metaphor for how the world is visualised today. We assemble the world from pieces, assuming that what we see is both coherent and equivalent to reality. Until we discover it is not."[17] We assemble world views and the world, and each assemblage calls for an assembly of bodies, techniques and sites.

The images that we have come to identify of the cosmos in past decades have often been taken by institutions that propose to shape our world view. Their images are hi-definition, scientifically verifiable, taken using the most advanced technologies of computation and specialised teams. The objects of their missions are unfathomably far and their objective is to discover what lies outside the already available knowledge of the universe and to capture it in high

16 Nasim, Omar, *Extending the Gaze: The Temporality of Astronomical Paperwork*. Science in Context, 26, 2013

17 Mirzoeff, Nicholas, ibid.

resolution. For an amateur astronomer the field of vision and the assembly of observers is guided by other parameters of discovery that explore and sustain a multiplicity of resolutions of world-making.

REFLECTION (Rohini): The amateur astronomer is not deterred by the poor image, but this is not the same as not wishing and often obsessing about being able to record a better one. The community of astro-photographers for instance, begin with what they have, but their journey often entails exhaustive research into equipment and media that will extend their gaze. Anna Henchman in her paper on the 'Telescope as a Prosthesis' describes the binary that is often set up, between the astronomer as a kind of seer who can see far and beyond and the machine which, as a kind of archetype of objectivity, transcends the fallibility of the human eye. She argues that this might be useful in understanding why some place greater importance on the astronomer, for whom the telescope is just another tool, while others might argue that the telescope could function without any human input at all.

"Any account of the role of telescopes in the history of astronomy tends, at least implicitly, to be based on some conception of the relationship between eye and telescope, and by extension, the relationship between human being and machine."[18]

She goes on to describe how John Herschel placed particular importance on 'practicing to see', on training the human eye and mind over the powers of the telescope.

REFLECTION (Sabih): The images of Saturn, Neptune, Uranus and the stars forming through the telescope that night almost felt delicate. It is hard to imagine that those giant planets would tremble with every little motion of your eye, and flicker almost like low-resolution gifs. After Saros 132, I am unable to see gifs the same way again. I am convinced that all the gifs in the world right now are possibly signals from millions of light years away. For the amateur astronomer, what the image cannot capture in resolution is captured in the resolution of collective practice. I am reminded of Hito Steyerl when she writes "(t)he poor image constructs anonymous global networks just as it creates a shared history."[19]

18 Henchman, Anna, "*The Telescope as Prosthesis.*" Victorian Review, vol. 35, no. 2, 2009, pp. 27–32. *JSTOR*
19 Steyerl, Hito, *In Defense of the Poor Image*, e-flux Journal #10, November 2009 (https://www.e-flux.com/journal/10/61362/in-defense-of-the-poor-image/)

If the practice of an amateur observer comes with the resolve to share, circulate and strengthen the resolution of observational practices, who defines the resolution of the universe?

Celestial Bodies

OBSERVER:
"When each one of us was looking through the telescope and checking what s/he was seeing by describing it in words, and between the 20 odd people it was as if we were conjuring that image. It made me think about what I saw from the telescope. How un-spectacular and bewitching it looked. How far away, how in looking at it the universe had to be acknowledged, and its thickness in between. The telescope helped us see something so far, but doesn't bring it close. That night wasn't just an event, but a thought incident."
Shveta Sarda, The Night Sky Observation, Alwar Rajasthan 2019

REFLECTION (Sabih): In April 2019, the first ever image of the black hole was released. Estimated to be 500 million trillion kilometres away, it was 'photographed' by a network of eight telescopes located in different parts of the world using a conjunction of mathematical, computational and visual techniques. From tiled renderings of images taken from a single satellite for over 8 hours, to images conjured by a network of satellites—spread across Hawaii, Mexico, Arizona, Atacama, Spanish Sierra Nevada and Antarctica scanning for over a period of 10 days—something has fundamentally changed the field of vision. Scientists describe imaging the black hole as a process that tries to construct an image by seeing through a shattered mirror.

Astronomers have pondered for generations what a black hole consists of, both outside and inside. While several answers have been proposed, a crucial detail that remains missing is its centre. The question is not to reinforce a centre in relation to a periphery. Rather, when something falls into a black hole and disappears, does it continue falling or does it reach somewhere?

Speculations suggest that perhaps at the centre of a black hole, everything melts into a cloud of probability - where space, time and matter no longer bear any meaning. Others suggest that black holes radiate heat because matter that enters them is transformed into heat.

Theoretical physicists like Carlo Rivelli, together with scientists at Grenoble and Nijmegen, have been exploring whether matter indeed slows down and stops before reaching a centre with an enormous pressure that prevents its collapse. Instead, they propose that matter stops falling and forms extremely small and dense stars called 'Planck stars', leading to something that inevitably happens to all matter in such cases—a rebound.

This rebound, like a ball bouncing back but in temporal reverse, transforms the black hole into its opposite, what they call a 'white hole'. A white hole, they claim, is a region of space into which nothing can enter but from which things only radiate—a reversal of a black hole so to speak. But staying with the black hole, it is ultimately no more than a star that collapses and rebounds in extremely slow motion. Recent scientists say that this was not possible in Einstein's theory because the latter did not take quantum effects into account.

When a sufficiently heavy star collapses in on itself, it produces a black hole and falls into it. A star with the dimensions of the Sun, i.e., thousands of times bigger than the earth, would produce a black hole with a diameter of only one and a half kilometres.

Consider a space of knowledge and thought that does not cast shadows. Consider the space of collective thinking as a celestial body entangled with other bodies, with moments of shimmer, changing densities and producing atmospheric lenses for collective transformations.

REFLECTION (Rohini): When something falls to a centre what does it leave behind? The 'event horizon' or 'surface' of a black hole prevents us ever 'seeing' into the centre. It was believed for a long time nothing could escape this horizon until Stephen Hawking discovered that the horizon in fact emits radiation (minute quantities, but enough to reduce the mass and energy of black holes) which is believed over time to cause black hole evaporation. He went on to argue that this radiation was free of information—it was just a random collection of photons, because the horizon holds nothing. The body—the star that formed the black hole—has collapsed into the singularity and left the horizon. This suggested that if you were to look at the final state of the black hole you would never be able to retrieve the information of what went into its formation, the initial conditions if you will. This flew in the face of everything physics tells us is fundamental, the theory of reversibility. Simply put, given the state of something, we can predict its state at earlier and later times. The fact that the Hawking Radiation produced at the horizon could give us no information about the black hole's initial state was called the black hole information paradox.

Recent developments in quantum gravity and quantum field theory have offered an infinitely stranger picture. Physicist and string theorist Suvrat Raju's work on black holes and on the black hole information paradox suggests that the information in the radiation is not as random as Hawking believed—that, if looked at closely, patterns become visible. But only if the observables or data points are looked at *in relation* to others and not in isolation. He describes how a better understanding of quantum statistical mechanics has demonstrated ways in which information can be preserved in delicate correlations between observables.[20]

Consider the quantum fire shimmering on the surface of a black hole, gradually causing its slow evaporation. Consider also the information the shimmer holds which can only be read in its correlations. Here the collective contains more information than the sum of the individuals.

REFLECTION (Sabih): An entire mass of a star gets concentrated into the scale of a single molecule. A ricochet quantum force kicks in, the star rebounds and begins to explode. For the star, only a few hundredths of a second have elapsed. But the dilation of time caused by the gravitational field of this explosion is so strong that when matter begins to re-emerge, tens of billions of years have already passed in the rest of the universe.

Consider the speed of darkness as this magnitude, taking place in the span of a hundredth of a second, happening millions of times over, while tens of billions of light years pass. To observe and imagine such unknowable magnitudes is the work of theory.

As Glissant would say, "[t]he abyss is also a projection of and a perspective into the unknown. Beyond its chasm we gamble on the unknown. We take sides in this game of the world."[21]

20 Suvrat, Raju, *Breakthrough Discoveries of 2020 - Black Hole Information Paradox*, https://www.youtube.com/watch?v=tQf5mMQ2Pwo&t=3529s

21 Glissant, Édouard, ibid.

ALL THAT APPEARS WILL DISAPPEAR[1]
Ivana Franke

1. Dissolving a Chair[2]

Contents of Consciousness

There is a chair onstage. It is empty.

> A chair is an object that forms and modifies space according to our bodies. Our bodies fit into it in a particular position so that our legs can rest from the strain of carrying our weight. Usually chairs have four legs, a seat, backrest and sometimes an armrest. Chairs are widely used by many humans on a daily basis. A large number of humans spend up to a third of their lives sitting on chairs.This is what we all knew before we saw this chair.

This chair has a black steel frame. Its backrest and seat are made of wood and painted blue. It is quite an ordinary chair, and it doesn't look very comfortable. It is approximately four meters away from me.

In the classrooms of my elementary school there were chairs of a similar shape. They weren't very comfortable either. Four years, five hours a day, five days a week. I sat on them for about 4000 hours. Almost half a year of my childhood.

1 Taken from the eko *Blessing for the Journey* by Roshi Egyoku Nakao

2 Parts of this text are adapted from the lecture-performance, *Chairs outside of human consciousness*, at Episōdo 00, *Afterglow*, Yokohama Triennale 2020, curated by Raqs Media Collective.

I remember having dreamed that they had grown fat, soft, round cushions.

These dreamed up pillowed chairs remind me of a recent experiment I did with a vision-scientist friend. We drank tea made of magic mushrooms—containing psilocybin, a hallucinogenic substance—to observe how our perception would change. The color of the chair in my kitchen became much more vivid. Each particle of its surface was glowing and vibrating. Those particles then spread into the space around the chair and started swirling, creating fractal-like shapes. They even made some sounds. Like singing birds.

(Multidimensional) States of Consciousness

The chair I am sitting on here and now, the chairs of my childhood that are remembered, the cushioned chairs that I dreamed of, or the vibrating and fractal-like blue chair I hallucinated are all appearances in my consciousness. Existing in various forms along different times and spaces, they appear in multiple dimensions of my conscious states.

> It is often argued that consciousness comes in degrees: ordinary waking awareness, dreaming and psychedelic states all appear as distinct levels of consciousness in a hierarchical order. There is even a measure of consciousness (PSI) which can be used to identify, evaluate and order how conscious a creature is, much like ordering humans by age or height.
>
> But the analysis of phenomenological, semantic, physiological, and functional dimensions of conscious states show that although the psychedelic state is distinct from the state of ordinary waking awareness, it is neither higher nor lower than the state of ordinary waking awareness. This poses a challenge to the unidimensional, level-based approach and supports a multidimensional account of conscious states, which allows us to better grasp what it is like to be in each of those various conscious states.[3]

Consciousness ought to be construed as multidimensional.

3 Olivia Carter and Tim Bayne. 'Dimensions of Consciousness and the Psychedelic State'. *Neuroscience of Consciousness* 2018(1): niy008. doi: 10.1093/nc/niy008.

Perception (Naïve Realism)

The chair on the stage is clearly distinct from my other mind's chairs, because *this chair is there, in front of me now*. Although, when I talk about this chair on this stage, I am inevitably talking about my private experience of seeing it, in the same way that I was talking about experiences of dreamed and remembered chairs.

> According to naïve realists the function of perception is to recover an accurate—'veridical'—representation of the external state of affairs. What we see out there are not our private imaginations, they are public entities: shared and verifiable.
>
> When electromagnetic waves bounce from the surface of this chair, the light patterns fall onto our retinas and are picked up by millions of photoreceptors. These photosensitive cells convert the light into electrical signals and our brains process information by their shape, movement, and feature analyzers. Thus, the chair appears in our minds—exactly as it is out there.
>
> Electromagnetic waves bouncing off of the chair's seat are within the range of around 470 nm, which means that the seat is blue and the majority of us see it as blue.

Controlled Hallucinations

I remember in 2015, when an image of a striped dress was posted online, with the question, "What Colors Are This Dress?" and it caused a viral sensation. Within hours, millions of people from every continent had voiced their opinions about their perceived color of #TheDress: many saw the dress as #whiteandgold, while others saw it as #blueandblack. There was no group whose opinion could have been explained away as being the result of some psychophysical disorder, even though there were many attempts to suggest this.

4 Christoph Witzel, Chris Racey, and J. Kevin O'Regan, 'The Most Reasonable Explanation of 'the Dress': Implicit Assumptions about Illumination,' *Journal of Vision* 17, no. 1 (February 2017). doi: https://doi.org/10.1167/17.2.1.

5 Jessica A. Collins, Ingrid R. Olson, 'Knowledge is Power: How Conceptual Knowledge Transforms Visual Cognition,' *Psychon Bull Rev* 21, no. 4: (August 2014): 843–860. doi: https://doi.org/10.3758/s13423-013-0564-3.

In fact, this viral event turned out to be the largest color perception experiment to date.

> Scientific findings[4] showed that cognitive penetration of perception by the unconscious assumptions—thoughts, knowledge and beliefs about the reality represented in the dress photo influence which colors observers perceive, in this case the assumptions about the illumination. Interactions between motivation and conceptual categories that we assign to what we see are pervasive in our understanding of the visual environment.[5]

> Perception is a *controlled hallucination*, in which the brain's hypotheses are continually reined in by sensory signals arriving from the world and the body.[6]

Experience (Micro-idealism)

Returning to the chair onstage, the experience of perceiving this chair is not different from the experience of seeing a chair under the influence of psychedelic substances, except that the appearance of the former aligns more with the physical reality.

> Perception is often misleading, and all one can rely on, in fact, are the measurements of observed entities' physical and chemical properties.
>
> But after having learned everything there is to know about color vision from a scientific perspective, the complete physical truth, while staying in a black-and-white room, after she exited into the colorful world, Mary, the protagonist of the thought experiment[7], realized that physicalism is false. She didn't yet know what it feels like to see a chair.
>
> It is in our minds that an object becomes something that we can relate to as we look at it, remember it, dream about it,

6 Anil K. Seth, 'Controlled Hallucinations and the Phenomenology of Presence,' in *Ivana Franke: Retreat into Darkness: Towards a Phenomenology of the Unknown*, eds. Katia Naie, Heike Catherina Mertens (Leipzig: Spector Books, 2018).

7 Amy Kind, Torin Alter, and Robert Howell, 'Mary and the Black-and-White Room,' in *Philosophy of Mind: 50 Puzzles, Paradoxes, and Thought Experiments* (Abingdon, UK: Routledge, 2024). doi: 10.4324/9781003179191.

> imagine it, interpret it, name it, think about it and use it. As it appears in our phenomenal consciousness. Phenomenal consciousness, or experience, is a medium in which the chair, and everything else, can only appear.

A physical entity becomes *real* to me as I experience it. When I experience perceiving the chair, I *realize* it there.

There are no chairs *outside* of human consciousness.

2. Dissolving the I (Waking up)

(So-called) Dreaming

This morning, when I woke up in a hotel room in Switzerland, I recalled a dream: I'm sitting, drowsy, at the noisy airport after a very long flight during a chain-travel to which I took my special red round pillow, but I had forgotten it at a previous location. A friend of mine, who would be arriving from there, took it for me. She is bringing it with her tomorrow.

I had in fact lost the pillow at the previous destination, so she won't be bringing it. I must have dreamed that. No ... luckily that didn't happen, I dreamed that I lost the pillow, and she is bringing it with her. I couldn't tell for sure. Did I dream that I took the pillow with me in the first place? I was trying to reconstruct the order of events, but I couldn't resolve it for what seemed a very long time. I thought I should call her, and I finally woke myself up in the hotel room. I stand up to check if the pillow was in the room, only to realize that I am still asleep in my bed at home.

Anxious to wake up and risk losing my favorite imaginary soft pillow for good.

> When Zhuang Zhou woke up from the dream in which he was a happy butterfly, he didn't know if he was Zhuang Zhou who dreamt that he was a butterfly, or if he was a butterfly dreaming that he is Zhuang Zhou, the story goes.[8]

Such transformations can happen.

8 Zhuangzi, 'The Butterfly Dream,' in *Zhuangzi: Basic Writings*, trans. Burton Watson (New York: Columbia University Press, 2003).

(So-called) Hallucinating

Occasionally when I wake up but am not yet fully clear to the day, I see images in my mind that are apparently not related to anything I have ever experienced. Like someone else's dreams, but in a waking state. Except that they don't feel unreal—it is like watching different scenes through a surveillance camera that has been placed at what, to me, are unknown locations.

A young girl riding a bicycle on the green meadow at dusk. A panda's face close up looking at the camera. People in uniforms with triangular hats, waiting. A pigeon with leopard-pattern feathers about to sing.

> Images of various types, that apparently have no connection with the person experiencing them, such as hypnagogic (experienced just before falling asleep) and hypnopompic (experienced just before waking), eidetic, pseudo hallucinatory, oneiric, daydreams, afterimages, images from sleep deprivation are not created by the person undergoing them, but rather they come from elsewhere.
>
> It is a field still largely unexplored, and it concerns 'the politics of images no less than what the French call the imaginaire of a given society'.[9]

(So-called) Waking

When I finally woke up and remembered the thread—I am I, in the room where I was yesterday, where I have lived for a few years now, it is Tuesday, there is a schedule, I'm late—I started thinking about which selves I should wear to fit today's schedule.

Morning work, the Grumpy one—slightly arrogant and determined to get things done; then for meeting family later in the day, the Friendly one—the (too) nice and empathetic one to get along with. I'd prefer also the Slow one at some point, which is the quiet and introverted version, but completely inappropriate when people are around. Hopefully they won't just take over. There is also a risk

9 Jimena Canales, 'Flash Force: A Visual history of Might, Right and Light,' in *Ivana Franke, Alexander Abbushi, Ida Mommenejad: Seeing with Eyes Closed*, eds. Elena Agudio, Ivana Franke (Berlin: Association of Neuroesthetics, 2011), 7.

10 Anne Böckler, Lukas Herrmann, Fynn-Mathis Trautwein, Tom Holmes, and Tania Singer, 'Know Thy Selves: Learning to Understand Oneself Increases the Ability to Understand Others,' *J Cogn Enhanc.* 1, no. 2 (May 16, 2017):197–209. doi: 10.1007/s41465-017-0023-6.

that the Defensive one may break in—righteous, easily offended and always ready to pick a fight. Especially if the Vulnerable one has been apparently harmed (who would instantly go into self-pity and easily start to cry).

> As a part of a social neuroscience research project—participants practiced perspective taking on their own inner states by learning to identify and classify different inner personalities—thinking, feeling and behaving differently, into categories like Worriers, Hurriers, Optimists, Helpers, Managers, Firefighters, Exiles, Protectors, etc.[10]
>
> They chose a recent situation that they had experienced, learned to de-identify from the character that lived the given situation, and take a bird's-eye perspective on themselves and their inner states to describe the situation from the perspective of another character, then another, then another ... Some identified up to twenty-three distinct subpersonalities, each of those obviously having different points of view on a situation and saw things differently.
>
> The experiment's conclusion: Learning to Understand Oneself Increases the Ability to Understand Others.

Neither perception nor non-perception

Sitting in a hall with my eyes closed, I hear a bird singing outside the window, in low-pitch sound with a squeaky twiddle.

On closer introspection, the fact is only that I hear the warbling sound. The bird is gone, because I may be hearing a recorded sound someone is playing on a smartphone, or some other sound. Perhaps a sound of a squeaking chair. Spatial location is gone—if it's not a bird singing outside, it could be a chair squeaking inside, or some other sound originating very close to me.

On closer introspection, sound is anyway in my mind. It is not in my head; my attention stretches and I am where the sound is. It's not far nor close, I am the sound.

On closer introspection, there is sound. There is no distinction in vibration, I lost the I.

On closer introspection, there is no sound.

See?

3. Dissolving the Interface

Disappearing Thing (Resonance of the Unforeseen)

It is as if the white noise has descended over the object, and I lost the signal to form the image.

> Disappearance by displacing uncovers the loud absence of an ordinary thing.
> Disappearance by hiding brings forth the memory of a fragile thing.
> Disappearance by annulling silences a monumental thing but keeps it in one's mind.
> Disappearance by dissolving erases a solid thing as one knew it from one's mind.

Disappearing Categories (Retreat into Darkness)

One time, when I entered a room, my whole visual world disappeared. There was no notion of space, the room could have been huge or tiny. The only phenomena I felt were gravity and breath. After a while, faint shapes slowly appeared—I could see them, but I could not recognize them. Then my thoughts enmeshed them with true fiction; in the absence of known objects, a fictional story emerged, evoked by these forms. They floated on their own, danced with me, each form dissolved into darkness and appeared at different places. As one drops the mental habit of quickly extracting relevant information from the already familiar perceptual categories, a narrative expands in unexpected directions. As an octopus or a frog, the universe, or God.

Disappearing Outside (Seeing with Eyes Closed)

Sitting on a chair in the middle of the room, in which the screen of numerous blinking lights covered the wall, I closed my eyes.

A shimmering of colors, patterns, shapes, or figures in motion appeared in my head, a stream of images: birds flying, people passing by,

11 Ida Momennejad, in *Distant Feeling*, eds. Irena Bekić and Ivana Franke (Zagreb: Zagreb City Libraries, 2012).

circles and spirals swirling ... Some of those were very far away, as if stretching to infinity; at some point they felt very close, as if my head was in a fishbowl. An aerial view, trees with naked branches, it looked as if something long forgotten was reappearing in front of me.

There is nothing besides white flashing lights out there.

Disappearing Image

Take a minute to look at the drawing on the previous pages (202–203). The drawing consists of a rectangular wave periodic gratings of different spatial duty cycles used regularly in scientific experiments to measure visual acuity, the resolution limit of the visual system; of black and white stripes with straight edges.[11] There are no colors on the page, there are neither waves nor curved edges, no overlaps. The image is two-dimensional.

Similar to the act of looking at the chair, light bounces off of the material surface, it reaches your retinas, and is processed by your brain, which makes some guesses about what is out there. Your brain's hypothesis should clearly be black and white stripes with straight edges are on those pages.

Seeing into Being (Limits of Perception Lab)

When the chair's surface begins to vibrate, I realize that the color I see is my own private perception of this color. You see something entirely different which I will never know. I encounter the thresholds of my phenomenal world.

> The way one perceives and makes sense of one's own experiences, reflects on them, and builds ideas and models of what 'the world' is, what is 'me,' and what lies 'in between' is contested and critical. Perception, far from being a universally neutral technology for being in the world and making sense of it, actually produces worlds and selves.[12] Perception is a porous membrane between beings and their contexts;

12 Elena Agudio, Ivana Franke et al., 'Your country of two dimensions is not spacious enough. Limits of Perception Lab,' exhibition handout (Berlin: Savvy Contemporary, 2020).

13 Sangheeta Menon, discussion during the online symposium 'Summer Solstice Invocations,' part of the project *Your country of two dimensions is not spacious enough. Limits of Perception Lab*, at Savvy Contemporary, Berlin, 2020.

it separates, but it also blurs boundaries. Cultural-specific mental frames are conditioned by expectations of our societies, while implicit biases based on cultural values that we associate with our perceptions define our social interactions.

Attention to the *invisible* (that which is present but cannot be seen), the *unseen* (that which is visible but evades our attention), and the *hallucinatory* (that which is seen but it is not there), *forgotten* and *imagined*, are all potential sites of transformation. They can render the mental process—continuously changing mental contents—visible as the definite factor in the formation of positionalities we inhabit as human beings in the world.

Too dark, too bright, too small, too big, too complex, too simple, too far, too close to be seen.

At such thresholds lies the potential for repositioning and reorienting ourselves, by expanding our sensitivity to both the visible and invisible, human and non-human, big and small, near and far away.

Rather than asking what is our place in the universe, a question, '*how is the universe placed in you?*' will then inspire us to redefine our own self, and our own ways of looking at experiences.[13]

Interweaving (Known and Unknown)

Whatever phenomena or absence of phenomena we encounter, our minds are filling them with stories. Stories that were passed on to us, in different colors we learned to recognize in our specific social environments. Some, familiar colors I see, I know which are those colors. Some I don't see, those colors are unknown to me. Somebody else knows and sees those, to me invisible colors. Unknown colors are woven into all stories. Your colors are in my story. And their colors are in your story. Our stories converge, glide without interruption merging one into another, flowing, unfolding, changing, disappearing. Sparkling.[14]

14 Nanedi Kgalalelo Mahashe

MAKING BETS WHILE KNOWING NOTHING

Raqs Media Collective

Gyan chausar, mandi chauhatte, sarat paasa saar
ya duniya mein rachi baazi, jeet bhavai haar
sadhu, sant, mahant, gyani, chalat kat purkaar
lal girdhar, jeevana din chyar

Knowledge is a game, dice roll, players rise and fall
But the world declares, "all bets are off," 'coz "victory is defeat"
Good lord, listen, be wise, your knowing days are numbered
—Mirabai, sixteenth-century princess, rebel, and mystic
(free translation)

1.

What would happen if one placed a bet on the idea that "knowing nothing," the knowledge of nothingness, is the basis of wisdom?

This is actually not a very new idea, even though it remains a surprising and unsettling one. Recognition of nothingness, or emptiness, or *shunyata*, is at the very heart of an ancient way of knowing and acting in the world. It expresses a gentle skepticism about three complementary proposals, familiar to us from both the natural sciences and some forms of theological doctrine—namely, the assertion that the universe is either one, totally knowable thing, or that the self is coincident with the cosmos, such that to be self-aware is to know the universe (the *aham brahmasmi* principle in nondualist Vedanta) or that actually everything is just random, pointless, and chaotic.

There is a fourth way—and that consists in knowing the primacy of nothingness as a binding principle. That a sense of emptiness, an awareness of the absence of substance and form, is itself generative of substance, form, and knowledge. In some traditions of knowing and acting in the Indic world (including classical Madhyamika Buddhism and some common traditions of knowledge and spiritual questioning), this would be called *nirakar* (formless) awareness.

The insistence on the primacy of nothingness is not an admission of ontological failure or epistemic collapse; rather it is a recognition of the fact that what binds one thing to another cannot be easily named, marked, or numbered. Its existence is signaled not by its substantiveness but by the way in which its massive immateriality affects other things.

2.

The Indian American theoretical physicist Subrahmanyan Chandrasekhar wrote about mysterious entities in his book *The Mathematical Theory of Black Holes*, published in 1983: "The black holes of nature are the most perfect macroscopic objects there are in the universe: the only elements in their construction are our concepts of space and time."[1]

The business of *creating objects whose elements are concepts* is a terse but elegant way of talking about what it is that some contemporary artists do. It may well be said that the relationship between contemporary art practice and the generation of new knowledge is not just a way of rendering information into aesthetically pleasing or challenging objects and processes. It is not just a matter of translating "research," or even deep archival investigation, or processing an encounter between different fields of knowledge to produce an artwork. Ultimately it boils down to a question of giving form to things that might be unthinkable without the act of giving form to them. It is the making of art, in this instance, that makes it possible to think a new thought as well as the other way round. Sometimes this may mean attempting to think the unthinkable.

What might the unthinkable be?

1 Subrahmanyan Chandrasekhar, *The Mathematical Theory of Black Holes* (New York: Oxford University Press, 1992), 1.

When Albert Einstein said that God does not play dice with the universe, he might as well have been objecting to a joyously skeptical song attributed to the Mewari Rajput princess, rebel, and mystic known as Mirabai, who lived, ecstatically, in the sixteenth century of the common era, in what is now Rajasthan, in western India. The words she sang, or so it is said, which serve as the epigraph to this essay, would have offered no comfort to Einstein.

In some ways, understanding what a black hole might be represents both the victory and the defeat of knowledge. If at least one element of knowledge consists of information, then a black hole, by its very nature, is the defeat of information.

So what would we be saying if we said that she knew something about a black hole?

If knowledge is a stab at omniscience, a claiming, a carving out, or conquest of cognitive territory—from the continent of bleak nescience—then how many moves must one make in the course of these maneuvers before victory and defeat begin to mirror each other at every step?

Is it possible, like Mirabai, to celebrate the risk of losing one's sense of self, or coherence, in order to be able to find joy in this dance, or even a sliver of rapture, in the playing of this game?

And no, the wager is not simply on ignorance as the threshold and foundation of all acts of knowledge. Rather it is the conjecture, and the paradox, that insists that to know anything at all, one has to at least know of that which is categorically unknowable. It's like realizing that there is a zero, a nothing, and a something, a one, that is hidden in every number. A nothing that adds, multiplies, divides, and a one, a singularity, that accumulates, and mates with itself, to make any quantity at all possible.

What does it mean to know that nothing?

What does it look like?

What does it sound like?

What does it feel like?

What does it taste like?

Mirabai, the *Namak Haram*, the wanton, the breaker of conjugal promises, spills her salt, she who no longer listens to her master, her husband, her king. She goes back on her word. So that in the beginning there is no longer a word, no leg left for her to stand on, even as she dances, makes her moves in the game. No words left to catalog her books. No title to her claims, no quarter. But a little bit of nothing lingers in everything, a zero inside every number, a silence in everything said. "Your knowing days are numbered."

But Mirabai's songs are sung forever.

How to share nothing? Not how not to share, but how to share that fullness of hunger, that foreboding, that foresight that comes with knowing more the taste of salt in your tears than stars in the sky or silence in the library.

3.

On July 31, 1930, Subrahmanyan Chandrasekhar—Chandra, to his friends and family—set sail from Bombay Harbor on the steamship *Pilsna* of the Lloyd Triestino Shipping Company, via Aden, Port Said, and Alexandria, for Venice, on his way to the port of Southampton in England. From Southampton, the nineteen-year-old Chandra was going up to Cambridge with hopes of studying astrophysics. He had three books to keep him company on the monthlong voyage, one of which was the celebrated British physicist Arthur Eddington's recent book *The Internal Constitution of the Stars.*

Eleven years prior to Chandra's voyage, on May 29, 1919, Eddington had found himself marooned on a ship anchored near the island of Principe, in the Gulf of Guinea, just off the west coast of the African continent. Eddington's mission was to scan the clear African sky in order to observe and measure the slight bending of starlight under the influence of the gravitational field of a massive object like the sun.

A total solar eclipse, clearly visible in the equatorial region of the Gulf of Guinea on May 29, 1919, constituted the perfect set of conditions for observing the very slight bending of starlight consequent to the approaching transit of the Hyades star cluster at the edge of the eclipsed, blacked-out disk of the sun.

Inspired by his quest, Eddington is thought to have joyously, and somewhat impishly, composed the following limerick,

Oh leave the wise our measures to collate
One thing at least is certain, light has weight
One thing is certain, and the rest debate—
Light-rays, when near the Sun, do not go straight.

If light did bend as it approached the sun, it would prove the empirical validity of the theoretical conjectures of the general theory of relativity, written up by the Bern patent clerk turned physicist Albert Einstein. On May 29, 1919, starlight bent as Einstein had said it should, measures were taken, and the hypothetical conditions laid down in the general theory of relativity calmly ascended to the pedestal of empirically verified truth. Then the world changed forever.

But there was a new problem. If the general theory of relativity, with its description of the bending and curving of space-time (which is what caused starlight to bend) under the gravitational pull and stretch of massive cosmic entities, was true, then it would also imply, as a corollary, the intensification of gravitational force to an extent that could tear holes in the fabric of space and time itself. Supermassive stars, no longer able to burn with their accustomed intensity, could collapse onto themselves under their own weight and become either dead, cold, frozen stars, called white dwarves, or even more strangely, turn into intensely dense singularities, capable of gravitational force so tremendous that they would swallow even light. The latter included the possibility of entities into which light, information, and matter would fall, seemingly to be lost forever. Relativity seemed to describe the way things worked, but it also invoked the conceptual probability of entities that seemed to defy reason and stretch the imagination. The implications of Einstein's theory played havoc with his desire for order.

The nineteen-year-old Chandra whiled away his monthlong sea voyage on board the *Pilsna* on mathematical calculations that accompanied his reading of Eddington's book on stars. These calculations suggested that according to quantum mechanical principles at least, the remaining energy of the atomic particles within a dying star could still act against gravity to keep it going as a white dwarf. Chandra's conjecture held out an important qualification. It proposed that this defiance of gravity in an imploding star had a limit. Any star more than 1.4 times the mass of our sun would collapse into itself to form a "singularity"—something that no one quite understood at the time—which would be supermassive but would lose almost all its matter in the form of a gigantic explosion—what we now call a supernova burst—that would distribute matter and leave behind an enigmatic residue.

The 1.4-times-the-mass-of-the-sun limit is now called the Chandrasekhar limit, in honor of the then nineteen-year-old.

Chandrasekhar worked on those shipboard calculations for the next five or so years and then wrote a brief paper titled "The Highly

Collapsed Configurations of a Stellar Mass."[2] The then twenty-five-year-old presented these conclusions at a meeting of the Royal Astronomical Society in London in January 1935. The paper contained the following tentative statement: "For all stars with a mass greater than M, the matter does not become degenerate. Great progress in the analysis of stellar structure is not possible unless we can answer the following fundamental question: what happens if we go on compressing the matter indefinitely?"

The answer to his question—"what happens if we go on compressing the matter indefinitely?"—would have to admit to the enigmatic possibility of black holes. Chandra's onetime hero Eddington, who was present at the meeting, was furious and put all his formidable rhetorical force into admonishing the young scholar, implying that Chandra's conclusion reduced the idea of relativistic degeneracy of stellar matter to an absurd level: "The star has to go on radiating and radiating and contracting and contracting until, I suppose, it gets down to a few km radius, when gravity becomes strong enough to hold in the radiation, and the star can at last find peace.... Various accidents may intervene to save a star, but I want more protection than that. I think there should be a law of Nature to prevent a star from behaving in this absurd way!"

The ridicule and humiliation that Chandra faced came out of incomprehension but was not without an element of racial prejudice and imperial arrogance. He decided to stop being an astrophysicist and went into the more abstract discipline of theoretical physics, left Britain, and did not write anything about dying stars for the next thirty-odd years. Four years after Chandra's talk at the Royal Astronomical Society, J. Robert Oppenheimer and H. Snyder published a paper titled "On Continued Gravitational Contraction," in which they proposed the possibility of the continued collapse of a star into smaller and yet more supermassive entities.[3] They were endorsing, at least in theory, what Chandra had implied.

Then in 1963 Roy Kerr, a New Zealander working at the University of Texas at Austin, published what he said was the result of observing an object rotating in a vacuum under conditions of general relativity.

2 Subrahmanyan Chandrasekhar, "The Highly Collapsed Configurations of a Stellar Mass," *Monthly Notices of the Royal Astronomical Society* 95 (January 1935): 207–25.

3 Robert Oppenheimer and H. Snyder, "On Continued Gravitational Contraction," *Physical Review* 55 (1939): 455–59.

These mysterious results were given a name, *black hole*, following the banter after this result.

The name *black hole* stuck to the nothing that it sought to describe.

4.

Something strange began to happen in 2015. We began to hear black holes.

On September 14, 2015, a gravitational wave designated GW150914 was recorded, and then rendered as a faint, 0.2-second-long "whump" or "chirp." It is actually possible to hear this sound, and it doesn't sound very remarkable.[4]

This wave beached on the shores of human consciousness when two black holes, one thirty-six times denser than the mass of our sun and the other twenty-nine times denser than the mass of our sun, spun into each other and merged to form a supermassive black hole that is estimated to be both appreciably smaller than, as well as roughly sixty-two times denser than, the mass of our sun.

The gravitational radiation that these two black holes released in the final fraction of a second of their merger is thought to be at least sixty times the total output of energy of all the stars in the observable universe. The actual event of the collision took place roughly 1.3 billion years ago, in the vicinity of the Magellanic Clouds (a pair of irregular dwarf galaxies) orbiting the Milky Way, in the southern celestial hemisphere.

The event (and ten others like it that followed, nine involving other black holes and one involving the collision of two neutron stars) was closely monitored by a network of one thousand scientists located in observatories spread across the world and by 440,000 lay enthusiasts, who were conducting observations from their personal computers, as a part of the Laser Interferometer Gravitational-Wave Observatory (LIGO)-VIRGO research process in the emerging field of gravitational wave astrophysics.

A black hole is a massive concentration of gravitational force, consisting of a mass condensed to such a great extent that it amounts to an "almost nothingness." Listening to the sound that we think two black holes make when they fall into each other is like being tuned in to the frequency of the biggest fragment of

4 The audio is available at https://www.gw-openscience.org/audio/.

"nothingness" that we have come across so far. It is the biggest nothing we have ever heard or have imagined hearing. And we have heard nothing like it before.

Then, on April 10, 2019, another array of instruments, and another network of scientists and enthusiasts, were able to generate the first recorded image of a supermassive black hole located in the Messier 87 (M87) galaxy, in the constellation Virgo, approximately fifty-five million light-years away from our earth.

It is estimated that this black hole may have eaten matter and energy equivalent to six billion suns. Looking into its image is like looking into the abyss of nothingness magnified to an unthinkable dimension. It is the biggest nothing that we have ever seen. And again, we have seen nothing like it before.

5.

What does it really mean to know "nothing"?

The phenomenon of a black hole plays havoc with our ordinary sense of time. Should a human observer fall into a black hole, she may emerge (at the hypothetical other end) in a parallel universe in which time runs contrary to the way we know it. The threshold of an event in a very remote past may also be the gateway to an unknown future, because of the way in which space-time folds around a massively dense concentration of the nothing that marks the presence of the black hole.

The birth of the black hole as a concept marks an event in human consciousness coincident with thinking the unthinkable, and its continued career can be represented by the evolution of unthinkability.

This order of knowledge that apprehends realities in a way that is impossible to describe in language, that is pre- or translinguistic, borders on the ineffable. This order of knowing is also a form of not-knowing, or a form of knowing emptiness, because it cannot be translated into information.

The Madhyamika Buddhist philosopher Dharmakirti developed a model of two truths—*satyadvaya*—as a description of the world in his text *Madyhamakavatara* (*Introduction to the Middle Way*). The first truth, *paramartha* (ultimate meaning), consisted of things that one could be aware of but that were impossible to contain within language; they could be said to be non- or preconceptual, of an order of awareness designated as *pragna*, a kind of wisdom that is beyond knowledge. The second, *samvriti*, or concealment, were conventional truths, the mirrors of our linguistic forms, concealed within the names we give to things. It could be said then that our apprehension of how matter and time behave when a supermassive celestial body crashes into itself is a movement from intuition and awareness to naming and observation and then again to a higher-order awareness of the unthinkable.

Considering a black hole, as we have seen, means considering the eventual loss of information. Knowing a black hole means knowing that knowledge itself dissolves once it crosses the black hole's event horizon. And yet, with it, we heard the inaudible and saw the invisible, and now, once again, we are beginning to think the unthinkable, anew.

This spiraling and back-and-forth movement between two different kinds of knowing, from *paramartha* to *samvriti* and then back to *paramartha*, is one way in which to create an image for the evolution of our consciousness.

In his lecture "Shakespeare, Newton, and Beethoven, or Patterns of Creativity," Subrahmanyan Chandrasekhar comes to a remarkable conclusion:

> In my entire scientific life, extending over forty-five years, the most shattering experience has been the realization that an exact solution of Einstein's equations of general relativity, discovered by the New Zealand mathematician, Roy Kerr, provides the *absolutely exact representation* of untold numbers of massive black holes that populate the universe. This "shuddering before the beautiful," this incredible fact that a discovery motivated by a search after the beautiful in mathematics should find its exact replica in Nature, persuades me to say that beauty is that to which the human mind responds at its deepest and most profound.[5]

6.

One reason to make art is because it condenses what happens in our minds when we feel compelled to rethink everything that we

5 Subrahmanyan Chandrasekhar, "The Nora and Edward Ryerson Lecture: Shakespeare, Newton, and Beethoven, or Patterns of Creativity" (1975), in *Truth and Beauty: Aesthetics and Motivations in Science* (Chicago: University of Chicago Press, 1987), 54.

know and take for granted. Sometimes this can begin with a meditation on a piece of sky. In a series called *Unfamiliar Jataka Tales*, we attempt something like that when we consider a patch of fallen sky. Being a lenticular text-image diptych, the image very appropriately takes advantage of the property of light to bend when refracted off a ridged surface chanced upon by a moving observer. In a moment of whimsy, it may well be regarded as a homespun, handheld, embodiment of the principles of general relativity in action. The work, like the conventional Jataka tales, tells a brief story about the career of an enlightened being, a bodhisattva. It merits retelling.

> How the most terrible solitude was overcome
> Once, a small rectangular piece of the sky (a Bodhisattva unknown even to itself) landed on the earth. It rested, bewildered by the hardness of the surface it found itself on. Everything, everywhere, was arid. The sky-patch-Buddha-to-be felt very alone, especially as it had lost its playmates, the clouds. Many years passed. The patch of sky, out of unself-conscious yet boundless compassion, decided to be useful to the earth on which it had found itself. The skylet, after considerable meditation on being fluid and substantial at the same time, condensed itself into a pool of clear, cool water. Now, grasses began to grow around it in abundance for miles on end. The wind played ripples and laughed on its surface. It quenched the thirst of passing strangers, provided clear water to splash about in on hot days. And its lost playmates, the clouds, would often stop by and reflect themselves. Sometimes it was just the clear blue firmament that found itself recovered in the empty blueness of the fragment of sky that had strayed so far from home. And so, this humble patch of fallen sky was united again with the heavens above, even as it continued to sustain the earth below. Grass, wind, clouds, earth, thirst and the quenching of thirst, all found themselves intricately bound in a relationship of inter-dependence. Nothing seemed possible any longer without the active presence of everything else. The desolation of the most terrible solitude was overcome, even as the most fulsome emptiness was attained.

Today, on rereading this story and looking again at the lenticular image of a patch of water on the earth that mirrors the sky, we are struck by a strange thought, a thought that conquers everything just as much as it is itself conquered by its own strangeness.

Could it be that this little patch of sky, or something like it, may have fallen into a supermassive black hole, and fallen out of it into a completely different universe, one like our own? Could it be that the making of a work of art is sometimes just an instance of information lost in one universe finding itself in another, and super-translating itself on the threshold of the event horizon of the consciousness of the artist? Even as one writes these words, their meaninglessness and utter absurdity become apparent. In the throw of the die of a concept, this is a move that means losing what one has, by way of reason, almost utterly. But then, nothing ventured, nothing lost, and nothing gained.

Is this what it might mean to shudder at the strange beauty of what it means to know, and know nothing, at the edge where art, science, philosophy, reason, and madness meet?

We could spend the rest of our lives trying to answer this question.

Matter Mattering Matters

Edited by
Kabelo Malatsie & Lantian Xie

Publishing Editor
Agnese Cantelmi (Mousse)

Copyediting & Proofreading
Kabelo Malatsie & Lantian Xie,
Liang Luscombe, Tracy Munirik

Translator
Khutatso Seshayi
Jazael Olguin Zapata

Design
Matteo Gualandris (Mousse)

Cover design
Simnikiwe Buhlungu

*Text excerpt sampled from footnote number 5 in *Tremblement de Terre*[1234567] from *besides Puleng; dontsa-ring and roving preoccupations*, co-published by Chisenhale Gallery, London, and Kunstinstituut Melly, Rotterdam.

Co-published by Kunsthalle Bern
and Mousse Publishing

Distributed by Mousse Publishing
Contrappunto s.r.l.
via Pier Candido Decembrio 28
20137, Milan–Italy
moussemagazine.it

ISBN
978-88-6749-650-1

€ 25 / $ 27

The editors wish to thank Alla Semenovskaya, Mrs. Mudzunga, Kokwani Gija, the South Easterly, the Black South Easterly, the North Westerly, Charlie's Fish Shop, Dora & Vedran, the fern that cooled the earth, Saharan dust storms, Rohit, Mo, Izabel, Malachi, Mochu, Dr Uhuru Phalafala, Keyanna Jones, Khensani Jurczok-de Klerk, Lindokuhle Nkosi, Kunsthalle Bern team, Ursula Wirz Stiftung, Thyssen-Bornemisza Art Contemporary, Duke University Press, Kunsthalle Dubai, National Institute for the Humanities & Social Sciences (South Africa).

KUNSTHALLE BERN